IDEAS INTO HARDWARE

A History of the Rocket Engine Test Facility at the NASA Glenn Research Center

Virginia P. Dawson

National Aeronautics and Space Administration
NASA Glen Research Center
Cleveland, Ohio
2004

This document was generated for the NASA Glenn Research Center, in accordance with a Memorandum of Agreement among the Federal Aviation Administration, National Aeronautics and Space Administration (NASA), The Ohio State Historic Preservation Officer, and the Advisory Council on Historic Preservation. The City of Cleveland's goal to expand the Cleveland Hopkins International Airport required the NASA Glenn Research Center's Rocket Engine Test Facility, located adjacent to the airport, to be removed before this expansion could be realized. To mitigate the removal of this registered National Historic Landmark, the National Park Service stipulated that the Rocket Engine Test Facility be documented to Level I standards of the Historic American Engineering Record (HAER). This history project was initiated to fulfill and supplement that requirement.

Produced by History Enterprises, Inc.
Cover and text design by Diana Dickson.

Published by Books Express Publishing
Copyright © Books Express, 2012
ISBN 978-1-78039-685-9

Books Express publications are available from all good retail and online booksellers. For publishing proposals and direct ordering please contact us at: info@books-express.com

Table of Contents

Introduction

D URING THE FIRST DAYS OF OCTOBER 1957, the Lewis Flight
Propulsion Laboratory (now the NASA Glenn Research Center)
in Cleveland prepared to host an event called the Triennial
Inspection of the National Advisory Committee for Aeronautics
(NACA). Inspections rotated among the three laboratories, located
in Langley, Virginia, Sunnyvale, California, and Cleveland, Ohio.
The inspections included tours and talks intended to demonstrate the
host laboratory's accomplishments to members of Congress, the aircraft industry, and the press. Because favorable impressions by these
official visitors often proved critical for the next year's appropria-
tions, rehearsals for the inspections were always tension-filled
affairs. The 1957 inspection was originally intended to feature the
laboratory's work on turbojet engines, but the launch of Sputnik
would precipitate a last-minute change in the program's focus.

As was customary, John Victory, the NACA's secretary, reviewed
the talks before the inspection. When rocket engineer Adelbert
Tischler mentioned how adding fluorine to the fuel of the Vanguard
rocket might give it sufficient thrust to achieve spaceflight, Victory
immediately ordered this remark deleted. Although the modest
Vanguard rocket had been chosen to loft a satellite as the American
contribution to the International Geophysical Year, Victory disap-
proved of any association with a program that smacked of "space-
cadet enthusiasm" identified with science fiction writers and crack-
pots.[1]

On Friday, 4 October the Soviet Union launched Sputnik, the world's first artificial satellite. Overnight the stigma associated with the word "space" vanished. Tischler and other members of the small rocket team at Lewis worked through the weekend to craft additional talks for the upcoming inspection to reflect the stunning news of the world's first satellite launch and their own contributions to rocketry. They had spent a decade pushing the development of high-energy fuels for missiles, but much of their work was highly classified. Before Sputnik few people outside the propulsion community knew of their innovative testing of high-energy rocket propellants.

The highlight of the inspection tour turned out to be a "stop" at the laboratory's new Rocket Engine Test Facility (RETF). The imposing structure, built into the side of a picturesque ravine, was so new in October 1957 that it was not yet fully operational. John Sloop addressed visitors seated on folding chairs in the test cell. Sloop headed the Rocket Branch of the Fuels and Combustion Division at Lewis. He had spent almost a decade of his career advocating rocket research both within the laboratory and to the Greater Cleveland community.

While the audience enjoyed a view of the autumn foliage gracing the woods on the other side of Abram Creek, Sloop explained the use of liquid propellants in missiles like Jupiter, Thor, Atlas, and Titan. These rockets used conventional kerosene-based rocket fuels. Sloop suggested that higher energy propellant combinations like hydrazine/fluorine, hydrogen/oxygen, hydrogen/fluorine, and hydrogen/ozone needed further investigation: "We are interested in these propellants because they can put higher speeds into a payload, thus giving longer range, or can give the same range with less propellant," he said. Sloop went so far as to suggest spaceflight. He invited the audience to consider the efficiency of using high-energy propellants to hurl a manned satellite glider into orbit above the Earth. Further research on the practical problems associated with the use of these fuels, he stressed, was urgently needed.[2]

After Sloop's talk, George Kinney, another member of the rocket branch staff, discussed the problems of designing and testing rocket

The $2.5 million Rocket Engine Test Facility (RETF), viewed from the Abram Creek valley shortly after construction, was completed in 1957. The RETF was an experimental facility, capable of testing sub-scale engines and components up to 20,000 pounds of thrust. NASA C-45652

injectors—that complex rocket component where fuel and oxidant are mixed several instants prior to combustion. Kinney betrayed the laboratory's predilection for the combination of hydrogen and fluorine. He explained that different injectors' designs produced spray patterns that affected combustion efficiency. Trying out injector configurations in the Rocket Engine Test Facility would allow engineers to isolate and study how a particular design interacted with other engine components, such as the combustion chamber and the nozzle. The culmination of the rocket presentations was a tour of the facility given by rocket engineer Edward Rothenberg. He said that previously, rocket testing at Lewis had been limited to small test models. The new Rocket Engine Test Facility would allow the NACA to investigate the problems of high-energy propellants in rocket engines that approached full scale.

Although the Lewis rocket group had originally thought in terms of missile applications, after Sputnik it shifted its focus to the design of engines and components for space launch vehicles. Walter Olson, one of the early leaders in fuels research, described his assessment of the rocket work at Lewis just prior to the setting up of NASA in 1958: "We had selected high-energy liquid propellants as our little corner of the rocket world to work on," he said, "and events had shown that they were not really needed for the ballistic military purpose, and that left us saying what good are they? And the only thing that they really appeared to be good for was, indeed, spaceflight."[3] The new Rocket Engine Test Facility would become an important tool for advancing the design of rocket engines. For more than thirty years it remained an experimental facility, dedicated to advancing the design of the nation's rocket engines, especially those fueled with cryogenic propellants like liquid hydrogen.

In physical layout, the Rocket Engine Test Facility was actually a complex of several buildings with the test cell at its heart. The test cell had a factory-like appearance. "All the pipes and pumps—all the things that made the test facility a test facility—was what dominated your view when you stood there," research engineer Ned Hannum remarked. "And the experiment was often hard to see. It was often something small and something obscure, and besides that—because we didn't test-flight hardware, we tested experimental hardware—it often didn't look like a rocket." Because rocket engines sometimes blew up during testing, the building housing the control room (Building 100) was located about a quarter of a mile away. Tank trucks called mobile dewars containing liquid propellants were parked on the hill above the test cell.

Static test stands like the RETF allowed engineers to isolate and test-fire a rocket engine independently of the body of the rocket. Test stands developed in the late 1930s and 1940s were an essential part of rocket engine development. They helped eliminate some of the enormous expense and waste associated with actual launches. In designing the Rocket Engine Test Facility at Lewis, the rocket team followed in the footsteps of Robert Goddard and other rocket pioneers. What made their venture different was that they intended their facility to be used for research to investigate the behavior of fuels and to improve engine components. This research would encourage innovation and anticipate problems encountered by rocket engine manufacturers.

By transforming ideas into hardware and testing them in the RETF, Lewis engineers contributed to a body of knowledge drawn upon by government and industry. The publication of test results, along with new ideas for the design of rocket components, provided the aerospace industry with new theoretical and practical approaches to engine design. The RETF was used to investigate general problems associated with the development of engine components, most often the injector and thrust chamber. The test article was carefully instrumented to provide test data that could later be analyzed by research engineers who prepared research reports for publication.

In 1957, the Rocket Engine Test Facility was regarded by the NACA as a state-of-the-art facility—the culmination of rocket test-

View of the test stand through the lighted open door of the test cell, also referred to as Building 202. The two men are standing on a cement apron where liquid hydrogen tanks would be installed in the spring of 1958. This photo appears to have been taken about the time of the NACA Triennial Inspection in October 1957.
NASA C-45924

ing and related activities begun at the Cleveland laboratory during World War II. By the mid-1950s, liquid hydrogen was already well established theoretically as an ideal rocket propellant. What was lacking was practical knowledge of storage, handling, and firing a hydrogen engine. "There was never a question as to liquid hydrogen being the absolute best fuel on earth," declared William Tomazic, a young engineer who had started at Lewis in 1953 after a year at Bell Aircraft in Buffalo.[4] Combustion of hydrogen produces high energy—a fantastic 54,000 British Thermal Units (BTUs) per pound compared to 18,000 BTUs per pound for kerosene-based fuels. The development of hydrogen-fueled rockets was considered extremely desirable, especially for upper-stage rockets like Centaur.

Built at a relatively modest cost of $2.5 million in the late 1950s, the RETF at that time was the largest facility for sea level testing of high-energy rocket propellants in the United States. The RETF's test stand could accommodate rocket engines that produced up to 20,000 pounds of thrust. A second test stand, referred to as Stand B, was added in 1984. Stand B increased the capabilities of the RETF by simulating the low gravity of a space environment. Test Stand C, added in 1991, also utilized the RETF's existing propellant feed systems, atmospheric vent, air, water, electrical, and data-recording systems. It was intended to test seal materials and bearings for liquid oxygen and liquid hydrogen pumps and other components. Due to a funding shortfall, Stand C never supported a test program. In 1995, in anticipation of the extension of a runway at nearby Cleveland-Hopkins International Airport, NASA terminated all testing in the RETF. It remained vacant in "inactive-mothballed" status until 2003 when the Rocket Engine Test Facility was razed.

Because of its important role in the development of a national expertise in the handling, storage, and firing of liquid hydrogen rockets, the RETF was designated a National Historic Landmark by the National Park Service on 3 October 1985, as part of its "Man in Space" theme study. Confidence that liquid hydrogen could be used safely in the upper stages of the Saturn rocket would later give the United States the edge over the Soviet Union in the race to the Moon.[5] To quote a prominent rocket expert from Marshall Space Flight Center, liquid hydrogen fuel is "one of the most momentous innovations in the history of rockets during the second half of the twentieth century."[6] Lewis engineers were not the first to test a liquid hydrogen rocket in the United States. That distinction belonged to the Aerojet Engineering Corporation in 1945. However, the sheer volume of papers published by the laboratory in the 1950s made Lewis Laboratory a clearinghouse for liquid hydrogen know-how and influenced decision-making at the highest levels of NASA during early planning for the space program. The RETF's close association with national expertise in the storage, handling, and firing of liquid hydrogen is the reason for the facility's landmark designation.

The RETF facilitated in-house government research, typical of the quality and originality of the work of engineers employed by the NACA. Fundamental research in the period before 1958 was intended to anticipate the development and manufacture of an actual engine by at least five years. The strong research culture of the NACA period made possible the progressive development of this unique facility. Its development was preceded by several smaller test cells in which engineers learned how to handle cryogenic fuels. Cryogenic liquids, such as hydrogen, helium, nitrogen, oxygen, and methane, boil at extremely low temperatures, making storage and handling a technical hurdle. During testing of these fuels, Lewis engineers tried out different design concepts, particularly those related to injectors and thrust chambers. The RETF was used for experimentation that yielded data useful to industry in the development and manufacture of their rocket engines.

Prior to the founding of NASA, government aeronautical engineers generally did not assist industry directly in development. The output of the rocket section was measured in research papers, as can be seen in the list of reports appended to this study. Contributions to the development of a specific rocket engine by industry were indirect—a fortuitous by-product of the laboratory's dissemination of its research through publication and participation in conferences.

In 1958, when the former NACA Lewis Flight Propulsion Laboratory became one of three NASA research centers, the laboratory's focus changed. The national furor raised by Sputnik prompted a shift in priorities. Although the laboratory retained its strong research culture, Cleveland engineers began to work more closely with rocket engine manufacturers as the country struggled to catch up with the Soviet Union's spectacular achievements in space. Looking back on NASA's early years, NASA's first administrator, T. Keith Glennan, admitted, "In truth, we lacked a rocket-powered launch vehicle that could come anywhere near the one possessed by the Soviets. And it would take years to achieve such a system, no matter how much money we spent."[7]

After NASA came into being in 1957, Lewis engineers worked closely with Pratt & Whitney to solve some development problems of the RL10 engine for the Centaur liquid hydrogen upper stage. In the 1960s, the RETF ran three shifts in an effort to help solve the problem of combustion instability in the F-1 and J-2 rocket engines, manufactured by the Rocketdyne Division of North American Aviation. At the same time, RETF engineers continued their more fundamental research on ablatives related to engine cooling. In the 1970s and 1980s, researchers turned to new problems associated with engine reusability. Low cycle thermal fatigue—the structural weakness of metals caused by cyclic exposure to extremely hot combustion gases—was one of the problems encountered in connection with the Space Shuttle Main Engine. In the 1990s, the RETF was used for tests directly related to the development of a low-cost rocket engine by the aerospace company Thompson Ramo-Wooldridge (TRW). However, even when the research focus narrowed to industry's more pressing problems, the strong research orientation of those involved in generating ideas and testing them in the RETF remained evident in the papers published by the group.

Notes

1. Adelbert O. Tischler taped telephone interview, 19 Nov. 2002. See also John Sloop, *Liquid Hydrogen as a Propulsion Fuel, 1945–1959* (Washington, DC: NASA Special Publication-4404, 1978), 90.

2. Texts of the October 1957 inspection talks can be found in John Sloop "Propulsion in the 1950s," misc. papers, NASA Historical Reference Collection, NASA History Office, Washington, DC. An early member of the American Rocket Society, Sloop became particularly known for his work on hydrogen/oxygen and hydrogen/fluorine rockets. In 1960 he became Technical Assistant to Abe Silverstein, at that time Director of NASA Space Flight Programs. Sloop later became Director of Propulsion and Power Generation, Office of Advanced Research and Technology. Sloop's book, whose full citation is given above, was an extremely valuable source for this study of the RETF.

3. Walter T. Olson Interview by John Sloop, 11 July 1974, 51, Walter T. Olson Biography file, NASA Historical Reference Collection, History Office, NASA Headquarters, Washington, DC.

4. William Tomazic interview, 22 Jan. 2002. All interviews by author, unless otherwise noted, took place at the NASA Glenn Research Center, Cleveland, OH.

5. Sloop, *Liquid Hydrogen,* 230-243; Virginia Dawson, *Engines and Innovation: Lewis Laboratory and American Propulsion Technology* (Washington, DC: NASA Special Publication-4306, 1991), 167; and Asif A. Siddiqi, *Challenge to Apollo: The Soviet Union and the Space Race, 1945–1974* (Washington, DC: NASA Special Publication-2000-4408), 317-18, 840. See also Virginia Dawson and Mark Bowles, *Taming Liquid Hydrogen: The Centaur Upper Stage Rocket, 1958–2002* (Washington, DC: NASA Special Publication-2004-4230).

6. Ernst Stuhlinger, "Enabling Technology for Space Transportation," *The Century of Space Science* (Dordrecht: Kluwer Academic Publishers, 2001), vol. 1, 73-4.

7. See Glennan's comments on this issue in J. D. Hunley, ed., *The Birth of NASA: The Diary of T. Keith Glennan* (Washington, DC: NASA Special Publication-4105, 1993), 23.

chapter 1

Carving Out a Niche

A<small>T THE</small> NACA L<small>EWIS</small> F<small>LIGHT</small> P<small>ROPULSION</small> L<small>ABORATORY</small> in Cleveland, the hazards of rocket testing kept the small rocket section isolated from the rest of the laboratory. "At times we felt like missionaries surrounded by jeering unbelievers," John Sloop recalled. "We had complete control over our facilities, at the far edge of the laboratory grounds, and over our operations, which was an exception to laboratory practice. In this environment we developed a great enthusiasm, a drive to excel, and a desire to show one and all the great potential of rocket propulsion."[1]

As the group gained confidence, they issued each other whimsical licenses that qualified them as a "first class rocketeer," presented papers at the American Rocket Society, and enjoyed a growing reputation among national rocket experts for their pioneering research on liquid hydrogen and other cryogenic propellants.

A 1941 graduate in electrical engineering from the University of Michigan, Sloop had spent the war years at the Cleveland laboratory on a narrow research problem—the fouling of spark plugs in aircraft piston engines. After the war, he was deeply chagrined to learn that the United States had lagged far behind the Germans in two important and exciting new areas of propulsion technology—turbojet engines and rockets. Sloop and a small group of engineers were determined never again to become enmeshed in narrow development problems. They wanted to tackle the problems of American rocketry in their widest possible context.

Sloop wrote:

> When the rocket group [in Cleveland] first got organized in 1945 and surveyed the field, it quickly became apparent that we had a lot of catching up to do. The German work was read with great interest. The publications of the prestigious Jet Propulsion Laboratory, the U.S. leader, became our textbooks. To make a contribution so late with so few, our leaders wisely directed that we work in lesser ploughed fields. That is why we concentrated on high-energy liquid rocket propellants, combustion, and cooling and left solid rockets to others. It has remained so to this day.[2]

The Cleveland researchers' focus on the development of high-energy fuels research grew out of the realization that they had limited resources and were latecomers to a field that had begun to take shape in the 1920s and 1930s. Nevertheless, Sloop and his young colleagues would leave their mark not only on the Lewis Laboratory, which later managed the Centaur liquid hydrogen rocket, but also win respect from the nation's propulsion community for their advanced research on rocket fuels.

ON THE SHOULDERS OF THE PIONEERS

Robert Goddard, a professor at Clark University in Worcester, Massachusetts, had launched the first liquid rocket (fueled with gasoline and liquid oxygen) on 16 March 1926 in Auburn, Massachusetts. Goddard had dreamed of flying into space ever since his youth when he read the serialization of H. G. Wells' *War of the Worlds*. He mentioned spaceflight in *A Method of Reaching Extreme Altitudes* in 1919.[3] Stung by ridicule of this idea by the press, he became obsessively secretive. His fear that others would steal credit for his innovations deprived American researchers of the benefit of his pioneering work.

Several years later in Germany, Hermann Oberth published *Die Rakete zu den Planetenraumen (The Rocket into Interplanetary*

From a safe distance, rocket pioneer Robert Goddard uses a telescope to observe a rocket mounted in a test stand. His left hand rests on the control panel with keys for firing, releasing, and stopping a rocket test, undated, probably the 1930s.
Great Images of NASA, http://grin.hq.nasa.gov. 74-H-1245

Space). Oberth's book provided a compendium of all that was known about rockets up to that time, sparking a rocket craze among German youths, who flocked to the 1929 movie *Frau im Monde (Woman on the Moon).* With Oberth serving as technical consultant, the movie featured a fanciful rocket with the Moon as its destination. Oberth's work was critical to the advance of rocketry in Europe. Taking his theoretical work as their starting point, a talented group that included Franz von Hoefft in Austria and Wernher von Braun in Germany founded amateur rocket societies. By 1929 the German Society for Space Travel (Verein für Raumschiffahrt, or VfR) had over 1,000 members.[4]

Possibly more realistic about the need for the deep pockets of the military, von Braun signed a contract with the German Army in 1932 to develop rockets at Kummersdorf, an aeronautical research laboratory near Berlin. As war approached, the Nazi government supported the construction of a secret rocket test facility at

Officials of the Army Ballistic Missile Agency (later Marshall Space Flight Center), 1956. Rocket pioneer Hermann Oberth, foreground; propulsion expert Ernst Stuhlinger, seated behind on left; Commanding Officer H. N. Toftoy, standing on left; Wernher von Braun, Director of Development Operations Division, seated right; Dr. Eberhard Rees, Deputy Director of Development Operations Division, standing right.
Great Images of NASA, http://grin.hq.nasa.gov. CC-417

Peenemünde on the Baltic Sea. At Peenemünde the Germans developed the A-series of rockets. They produced the A-4, later renamed the V-2 (Vengeance Weapon number 2), the world's first operational missile. The V-2 burned alcohol with liquid oxygen.[5] After the war, von Braun and other Germans who worked on the V-2 would be recruited by the United States Army to become the nucleus of the Army Ballistic Missile Agency.

American efforts lacked comparable popular and military support, though amateur societies were organized in several American cities. The most prominent of these societies, the American Interplanetary Society in New York City, grew out of the enthusiasm of a group made up largely of science fiction writers in 1930. The Society struggled to raise funds for a series of static rocket tests, and, in an effort to distance itself from what was then regarded as the fantasy of interplanetary travel, changed its name to the American Rocket Society in 1934. The group's most important achievement in the prewar period was the design and successful test by James Wyld in 1938 of a regeneratively cooled rocket engine. This engine became the basis for Reaction Motors, Inc., of Pompton Plains, New Jersey, founded in 1941. As historian Frank Winter has written, "Now, liquid rockets could be adequately cooled so they could be fired over reasonably long durations instead of prematurely burning out due to overheating. This made the liquid fuel rocket a practicable engine."[6]

In Cleveland, a German engineer, Ernst Lobell, infected by the European rocket craze, helped to found the Cleveland Rocket Society, one of the most active of the smaller American rocket societies. Edward L. Hanna, grandson of the fabulously wealthy tycoon Marcus A. Hanna, helped to finance the development of two rocket motors, built to Lobell's specifications. Lobell aimed at nothing less than piloted flight into the stratosphere. The exorbitant cost of these endeavors during the Depression, and the loss of its moving spirit when Lobell relocated to another city, were too great for the fledgling amateur society. It quietly disbanded after exhibiting a rocket motor at the Paris International Exhibition in 1937.[7] When the

Theodore von Kármán (center) sketches on the wing of an airplane while the JATO team at Aerojet Engineering Corporation watches attentively. Left to right: Clark B. Millikan, Martin Summerfield, von Kármán, Frank Malina and Homer Boushey, pilot of the first JATO-equipped American airplane.
Great Images of NASA, http://grin.hq.nasa.gov. Ames JATO-VONKARMAN

Cleveland laboratory came into existence during World War II these brave experiments had been all but forgotten.

Meanwhile, on the west coast in the 1930s, Frank Malina, a graduate student at the California Institute of Technology, convinced his advisor, Professor Theodore von Kármán, to allow him to write a thesis on rocket propulsion. He recalled how his astrophysics professor Fritz Zwicky had advised him not to waste his time, for "I must realize that a rocket could not operate in space as it required the atmosphere to push against to provide thrust!"[8]

This comment expressed a common misunderstanding. Rockets do not push against anything. Their flight skyward is based on Newton's third law of motion—that for every action there is an equal and opposite reaction. Hot gases rush out the flared back end, called the nozzle, creating a force that propels the rocket forward. It is the same force that kicks the barrel of a rifle back against a hunter's shoulder when a shot is fired, pushes a driver back into the seat of a car when he or she steps on the accelerator, or propels an inflated balloon in erratic circles as air is expelled from its open end.[9]

During World War II, the success of the German V-2 as a psychological weapon to terrorize the British people awakened the U. S. government to the potential of rockets in the arsenal of democracy. The increasing sophistication of the Malina group's experiments in the Arroyo Seco above Pasadena's Devil's Gate Dam enabled them to win an Army contract for small "jet" assisted takeoff devices (called JATOs). JATOs were little rockets used for additional takeoff power for heavily loaded military planes and to shorten takeoff from island airfields in the Pacific.[10]

MISSIONARIES AND UNBELIEVERS

Late in the war, engineers at the Aircraft Engine Research Laboratory in Cleveland began to focus on the development of rock-

Preparing jet assisted take-off units, called JATOs, for testing in one of the four rocket test cells built during World War II at the Aircraft Engine Research Laboratory in Cleveland, Ohio.
NACA C-1946-14482

Jerome Hunsaker, chairman of the National Advisory for Aeronautics Main Committee (left), Vannevar Bush, wartime head of the Office of Scientific Research and Development (center), and George Lewis, NACA Director of Research, visit the Aircraft Engine Research Laboratory, 9 October 1946.
NACA C-1946-10241

ets for the first time. In 1944 they constructed four garage-sized rocket test cells made of cinder blocks. These cells could test small rockets of up to 100 pounds of thrust. Though this work was never sanctioned by a formal research authorization, it appears to have been carried on with the tacit approval of the laboratory's director.

After the war, the zeal of Sloop's small rocket group contrasted with the attitudes of NACA officials in Washington. Well past his prime at the end of the war, George W. Lewis, the NACA's Director of Aeronautical Research, and Jerome Hunsaker, Chairman of the NACA's Main Committee, considered rockets artillery and therefore outside the mission of the federal agency to promote aeronautical research. Vannevar Bush, former chairman of the Office of Scientific Research and Development, also took a dim view of investing national resources in the development of rocket propulsion. As late as 1949, Bush asserted that the astronomical costs of developing intercontinental ballistic missiles could never be justified because rockets were "already near the limit of the amount of energy that can be chemically packed into a given weight."[11] In contrast to the conservatism of NACA Headquarters, rockets caught the imagination and enthusiasm of the engineering staff of the Cleveland laboratory, renamed the Lewis Flight Propulsion Laboratory in 1948.

Walter T. Olson, named Chief of the Combustion Branch within the Fuels and Thermodynamics Division in 1945, encouraged members of the rocket section. They studied the papers by such Jet Propulsion Laboratory luminaries as Frank Malina, Martin Summerfield, and Richard Canfield.[12] One might speculate that a paper of particular interest to the Lewis rocket researchers could have been "The Problem of Escape from the Earth by Rocket" by Malina and Summerfield. This paper expanded upon Konstantin Tsiolkovskiy's idea that stepped rockets (rockets with more than one stage) would make it possible to achieve spaceflight. This was not the first study of this subject. As early as 1919 Goddard had speculated on the potential of multi-stage rockets in *Method of Reaching Extreme Altitudes*. The paper's timeliness, along with the precision and persuasiveness of its arguments, has made it one of the classics in American rocket literature.[13]

Though still in his thirties, Walter Olson was one of the laboratory's outstanding leaders. He was among the handful of engineers at Lewis with a Ph.D. in chemistry from the Case Institute of Technology. Drafted by the Army, he was assigned to the NACA, where he went to work developing lubricants and fuel additives for aircraft piston engines. During a NACA-sponsored recruiting trip to California, he had made a detour to Pasadena's Guggenheim Laboratory to visit Malina. Witnessing the ear-splitting tests in the Arroyo Seco Canyon proved a revelation. Upon his return to Lewis, Olson joined the American Rocket Society (ARS), where he was soon rubbing shoulders with Malina and Summerfield, Wernher von Braun, Krafft Ehricke, and other members of the former Peenemünde group. Recruited by the U. S. Army shortly after the fall of Germany to the Allies, the Germans joined the ARS, giving the society the credibility and expertise it had lacked in the 1930s.

Unlike today's large bureaucratic space agency, NACA engineers had considerable autonomy. All three NACA laboratories supported strong research cultures that nurtured individuals motivated by the desire to contribute to the nation's store of technical knowledge. Moreover, unlike NASA today, the NACA did not engage in development or support missions through the management of contracts with industry. As James Hansen, the historian of the NACA Langley Laboratory in Hampton, Virginia, has pointed out, the physical isolation of Langley researchers and their remoteness from Headquarters freed them from political pressure and allowed them to concentrate on technical questions.[14] Lewis rocket researchers shared this isolation and reveled in the opportunity to delve into the field of rocket propulsion. Though there were no facilities for expensive full-scale rocket tests, they succeeded in ferreting out valuable lines of inquiry. Armed with pencil and slide rule, Paul Ordin and Riley Miller made theoretical evaluations of the performance of different combinations of high-energy propellants based on their specific impulse.[15] Specific impulse (pounds of thrust per pound of propellant consumed per second) is a means of measuring the efficiency of a particular propellant combination. After the publication of this paper, co-workers called Ordin "Mr. Rocket Propellant."

Vearl Huff inspired one of the early research reports that caught the attention of the rocket propulsion community. With Sanford Gordon and Virginia Morrell, he developed a basic calculation technique that saved considerable time over other methods of evaluating the theoretical performance of different propellant combinations.[16] Morrell, who had a degree in mathematics from the University of Wisconsin, recalled that Huff "had more ideas within an hour than most people get within their lifetime." Huff handed her a problem with 25 equations and 24 unknowns. She set about solving the problem with the help of Marge Terry, on loan from the "calculator unit." (Members of this all-woman unit were generally referred to as "computers.") Using one of the early electromechanical calculators made by the Friden Calculator Company, Morrell and Terry worked through reams of data, applying pressure, volume, and temperature variables to arrive at a method to evaluate a rocket's thrust, referred to as a "rapidly convergent successive approximation process."[17] The method received considerable notice because it could be programmed on the simple analog computers in use at that time, and could be used as a guide for propellant selection before a particular combination was actually tested. In describing the value of this early theoretical work, Sloop wrote:

> This marked the beginning of a long series of published and unpublished theoretical calculations by Huff, Sanford Gordon, and other associates that guided us and others in experimental work and in propellant selections. By the end of the 1950's they had published some 25 reports. Equally important, they were ready to supply reams of tabulated machine calculations to analysts and experimenters. Thus, it can be said that theoretical performance calculation techniques and results were the first major Lewis contribution to rocket propulsion research.[18]

The 1951 report, called "General Method and Thermodynamic Tables for Computation of Chemical Reactions," was published shortly after Morrell left the laboratory to have her first child. She recalled that when her husband, Gerald Morrell (also a member of the rocket group and later division head), attended a rocket meeting in California, he was besieged by researchers who mistook him for the author. He came home considerably deflated and demanded a copy of the report.[19] This report became one of the classics of the rocket world, cited by Martin Summerfield in his article on the liquid propellant rocket engine in the landmark multi-volume survey, *High Speed Aerodynamics and Jet Propulsion.*[20]

The rocket group benefited not only from the encouragement of their branch chief, but also from his recognition of the importance of building bridges between theoretical approaches and the development of actual hardware. "There is always an interesting interplay when engineers are across the table from physicists," Olson commented. "Those were the days [the 1950s] where there were fairly sharp distinctions between pure research, which physicists and chemists did, and applied research, which engineers did. Often one group would decry the efforts of the other group. I think one of the things that I contributed was pushing those two types together."[21]

The interaction of scientists and engineers produced innovative approaches to rocket engine design and the development of increasingly sophisticated rocket test facilities. Ideas found expression in the design of test hardware. The focus of their effort was to understand the processes that were going on inside a rocket thrust chamber. Testing played a key role in the generation of new knowledge that could be disseminated throughout the entire propulsion community—a community that included other NACA centers, the military, and industry.

THE ALLURE OF CRYOGENIC FUELS

Initially rocket researchers at Lewis did not focus on hydrogen, though hydrogen's potential as a high-energy rocket propellant was well known. The 19th-century Russian rocket pioneer Tsiolkovskiy had discussed its allure.[22] So had Goddard and Oberth, but its well-known dangers had dissuaded them. At Kummersdorf in the 1930s, Walter Thiel, a member of the German Army group working under Walter Dornberger, had briefly experimented with it, but results were discouraging.[23]

Engineers at Lewis Laboratory knew of the efforts by the Aerojet Engineering Corporation in Azusa, California, and the Jet Propulsion Laboratory in Pasadena to test a hydrogen/oxygen rocket in the late 1940s.[24] They were also familiar with experiments at Ohio State University by world authority on liquid hydrogen,

Professor Herrick L. Johnston. By 1951 both programs had been terminated.[25]

Thus, at the very time when other more experienced rocket experts were abandoning cryogenic fuels research because of its formidable challenges, the Lewis team began to systematically explore the full range of liquid rocket propellants, the better to evaluate them. They experimented not only with liquid hydrogen, but also with hydrazine, liquid ammonia, and lithium. They tested them with an equally wide range of oxidizers like hydrogen peroxide, chlorine trifluoride, liquid oxygen, nitrogen tetroxide, liquid fluorine, and ozone. Usually, they could obtain chemicals only in small quantities. They took great risks to bring them safely back to the laboratory. For example, because hydrazine was too hazardous to be shipped, Paul Ordin purchased enough in St. Louis to be stashed in an inconspicuous flask for the train ride back to Cleveland.[26]

Fluorine seemed particularly tantalizing. The Manhattan Project to develop the atom bomb had stimulated interest in a uranium/fluorine compound, making fluorine appear a promising area of postwar research. Fluorine, however, was another highly reactive chemical. The Harshaw Chemical Company of Cleveland considered it so dangerous that it required a police escort in the dead of the night. On one occasion during unloading, a fluorine bottle rolled off the truck. It hit the ground and "took off like a rocket," veteran researcher William Rowe recalled—fortunately in a direction opposite from where he and others were standing.[27] Fluorine gas had to be condensed before it could be loaded into a storage tank. Later the laboratory was able to obtain liquid fluorine from the Allied Chemical Company.

Diborane also received close scrutiny. Its propensity to explode made it particularly dangerous to handle. Rowe and a colleague purchased a pound of diborane for $400 at the Buffalo Electrochemical Company. They carefully packed it in dry ice and put it in the back of their pickup truck. On the way back to the laboratory, the pickup's fuel line froze, and the pair stopped at a repair shop to thaw out the line. When Rowe noticed an open flame from a heater in the shop, he became extremely concerned because he knew that even a small diborane leak could blow the roof off the shop. Fortunately, they made it back to the laboratory without further incident.[28] Experimental performance of these propellants was evaluated through tests in the small rocket test cells built during World War II.

Gradually, the work of the rocket group came to the attention of the Navy Bureau of Aeronautics (BuAer) at the Naval Rocket Laboratory and the Air Force at Wright Field. In May 1948, the NACA sponsored a classified conference on rocket fuels in which Sloop, Ordin, and Huff discussed the theoretical performance of diborane with liquid oxygen, hydrogen peroxide, and liquid fluorine. They were also able to supply experimental data on the firing of diborane with liquid oxygen in a 100-pound thrust rocket engine.[29] Shortly after this conference, BuAer asked the rocket group at Lewis to study rocket ignition at high altitudes in connection with jet-assisted takeoff for fighter aircraft. This research was considered urgent because the British had discovered that liquid propellants ignited spontaneously at sea level, but encountered serious ignition problems at high altitudes. The Navy was particularly interested in the behavior of liquid oxygen/alcohol propellant combinations at altitudes between 50,000 and 100,000 feet. This became the first officially sanctioned rocket research: Research Authorization E-229, "Investigation of Rocket Ignition by Lewis Flight Propulsion Laboratory at High Altitude," issued in 1949. In 1950 the Navy asked the laboratory to investigate the low-temperature starting characteristics of an Aerojet rocket engine that burned white fuming nitric acid and gasoline with hydrazine hydrate used for ignition.[30]

CONTRIBUTIONS AND RECOGNITION

The first research authorization for NACA rocket research in Cleveland coincided with the promotion of Abe Silverstein to Chief of Research in 1949. He reorganized the laboratory, making Fuels and

Combustion into a separate division with Walter Olson in charge. The Combustion Branch within this division had about 40 people— 8 in the Combustion Fundamentals Section and 19 in the Jet-Engine Combustion Section. At this time the miniscule Rocket Section increased from a staff of 5 to 13.[31] With the total number of employees at the laboratory more than 2,000, this still amounted to a very small effort. Nevertheless, Silverstein, a Langley transplant who understood the value of research, encouraged the work of the group.

Silverstein was known for his infallible engineering instincts and ability to pick promising lines of inquiry. After his 1929 graduation from the Rose Polytechnic Institute (Rose Hulman Institute of Technology) in Terre Haute, Indiana, he had launched a distinguished career at Langley where he became an expert in wind tunnel design and operations. In 1943, Silverstein was placed in charge of engine testing in the Altitude Wind Tunnel in Cleveland. This facility, at that time the country's most advanced wind tunnel, played an important role in testing its first British and American turbojet engines.

Silverstein sensed the coming importance of rocket technology. He authorized the construction of four additional rocket test cells. These test cells, located at the far end of the laboratory, were constructed by an in-house construction group called the "Hurry Up Construction Company" and paid for out of operating funds.[32] Between the four older test cells and the four new ones, they constructed a small instrument and control building protected from the test cells by earthen mounds. They could now test rocket engines producing 1,000 pounds of thrust.

In May 1949, Air Force Colonel R. J. Minty, Chief of the Wright Field Power Plant Division, asked the NACA for more experimental data on high-energy propellants in relation to combustion chamber operating pressures. Presumably referring to the Cleveland laboratory's work for the Navy, he wrote, "The NACA program investigating special rocket propellant combinations has been followed with considerable interest. As an aid to formulating plans for future rocket development, the type of work now being done at Cleveland is of importance." Citing a lack of information on chamber pressures, he requested data on specific impulses obtainable at 1,000 and 2,000 psia with three propellant combinations: white fuming nitric acid/gasoline, liquid hydrogen/liquid oxygen, and liquid hydrogen/liquid fluorine.[33] The laboratory responded that its experimental program on rocket engines did not include investigation of the effect of high chamber pressures on performance. It asked for and received authorization from the NACA Subcommittee on Combustion to develop an altitude tank for testing small rocket engines.

At the same time, the rocket section began to formulate long-range plans. An outline of the objectives of the Combustion Division, completed in September 1949, described its rocket work as both fundamental and applied. It defined fundamental research as "exploratory and thus long range," such as understanding the physics and chemistry of the combustion process itself. Applied research involved the effort to apply this understanding to the operation of full-scale rocket engines. In describing the laboratory's rocket research, the document stated: "Because the component parts of the rocket engine—propellants, injection system, combustion chamber, and nozzle—are so closely interrelated, research directly pertinent to this power plant is outlined and conducted as a body, rather than under other headings."[34] These same interests would later inspire the design of the RETF.

The work at Lewis on liquid-fueled rockets focused on four objectives: first, to obtain high specific impulse (more energetic propulsion per unit mass of propellant) in order to increase range and/or payload; second, to increase operating time and reliability of operation; third, to achieve versatility, ease, and safety of operation and control; and fourth, to determine the limits of flight performance achievable either with the rocket engine alone or in combination with other power plants. Interestingly, in evaluating different propellants at this time, the laboratory decided (at least temporarily) not to include liquid hydrogen/liquid oxygen in the test program because it was reluctant to compete with larger experimental programs at Aerojet, JPL, and Ohio State University. Rather, rocket researchers focused on liquid hydrogen with the oxidizer fluorine because of fluorine's theoretical superiority to liquid oxygen.[35] Soon, they would have the field of cryogenic rocket fuels research to themselves, since by the early 1950s the Air Force and Navy had ceased to fund early efforts at Aerojet, JPL, and Ohio State.

One of the most recalcitrant problems in early rocket development concerned unstable combustion. In his memoir, *The Wind and Beyond,* Theodore von Kármán recalled that the Malina group, working in the early 1940s, had despaired of solving the problem. Either the rocket would blow up or the flame would go out. After the war, they discovered that the Germans had also experienced the problem. Martin Summerfield theorized that instability in the engine depended on the delay between injection of the fuel and combustion. If the time was almost instantaneous, combustion was likely to be smooth. However, commented von Kármán, "unfortunately we knew of no way to reduce the reaction delay to solve our practical problem."[36]

At that time researchers defined two types of combustion instability. One involved a phenomenon that produced low-frequency oscillations known as "chugging." The other type, called "screech," occurred in the high-frequency range. Because the physics of combustion are so complex, engineering science at that time lacked a theory to fully explain the causes of combustion instability. This problem at the nexus between theory and experiment attracted the Lewis researchers. At first they focused on chugging. Through experimentation, they demonstrated that by decreasing the propellant mixing time and increasing injection pressure drop, the problem could be controlled, if not eliminated. They presented these test results at a symposium at the Dover, New Jersey, Naval Air Rocket Test Station in December 1950.[37]

Lewis researchers devised new experimental approaches to the problem. In a paper presented to the American Rocket Society,

Marcus F. Heidmann and Jack C. Humphrey advanced a theory that fluctuations in the injector spray could explain screech. Judging from the published comments to the paper, the theory was controversial.[38] In another paper, Donald Bellman described how he used a camera that photographed combustion at the rate of 40,000 frames per second to study high-frequency oscillations. Theodore Male, William R. Kerslake, and Adelbert O. Tischler continued this work. They applied the photographic techniques of Cearcy D. Miller, a NACA pioneer in high-speed photography. During World War II, Miller had devised his ingenious camera to photograph the phenomenon of "knock" (incomplete combustion) in aircraft piston engines using a transparent cylinder. With high-speed photography, rocket researchers were able to show the transition from normal to oscillatory combustion and from longitudinal waves to high-speed rotary waves in a 1,000-pound thrust engine.[39] These findings were reported at another symposium at the Naval Air Rocket Test Station in 1954 and published the same year.[40] Much of this work, however, remained classified because of its missile applications.

A document summarizing the laboratory's research on high-energy propellants for long-range missiles stated:

> The first objective of research on high-energy propellants for long-range missiles is to provide information for the selection of promising propellant combinations. This is done by a combination of analyses and experiments to reveal propellant characteristics, handling methods, theoretical performance, and burning characteristics. Such work has been the primary emphasis of rocket research at the laboratory with results that narrow propellant selection to fluorine, ozone, and oxygen as oxidants and hydrocarbons, ammonia, and hydrogen as fuels but keeping an open-minded attitude for other possibilities.
>
> The second objective of research on high-energy propellants is to provide information on engine performance, durability, pumping systems, and controls on engines of practical size. Work on

phases of engine performance (injection, combustion, expansion), durability (heat rejection, cooling), and pumping are starting.[41]

At this time, Peenemünde rocket veteran Krafft Ehricke at the Guided Missile Development Group at the Redstone Arsenal in Huntsville, Alabama, began to champion liquid hydrogen as a propellant for upper-stage rockets. He told John Sloop in a 1974 interview, "I had run into hydrogen, from Tsiolkovskiy through von Hoefft to Oberth to Thiel and through my nuclear investigation. So I said, 'It's too often that it has looked good. I think we ought to do it.'"[42] Ehricke wrote a paper comparing the weight of propellants, their specific impulse and their density. In it he argued that while it was preferable to use heavy- or medium-weight propellants in the first or booster stage, less dense propellants like ozone/methane, hydrogen/oxygen and oxygen/hydrazine seemed attractive for upper stages. He noted how the high specific impulse and low density of liquid hydrogen would offset the weight of the structures needed to contain it.[43] But von Braun gave Ehricke little encouragement. He thought they should stick with the denser propellants that they already knew how to handle.[44]

Liquid hydrogen is a cryogenic fuel, which means that it must be maintained at an extremely low temperature (minus 423 degrees Fahrenheit) to prevent it from vaporizing. Liquefaction of gaseous hydrogen is both complicated and expensive, while storage of large quantities of liquid hydrogen presents another set of technical problems. Its low density requires relatively large insulated metal containers. Because of the extremely low temperature of liquid hydrogen, these containers over time become brittle and develop cracks. Hydrogen tends to leak through the pores of even the most skillfully welded vessels. Despite these drawbacks, liquid hydrogen looked promising not only because of its high specific impulse, but also because of its extremely cold temperature. Its thermal properties made it excellent for cooling by pumping the fuel through hollow passages in the thrust chamber walls. Called regenerative cooling, this process allows the fuel to cool the thrust chamber walls during combustion, plus prewarm the fuel before it is injected into the chamber, enhancing the efficiency of the combustion process. Regenerative cooling also provided the key to liquefying hydrogen in the late 19th century.[45]

James Dewar, a Scottish physicist and chemist, converted hydrogen gas to a liquid for the first time in 1898. He used liquid air to precool the hydrogen gas, which he then expanded through a valve in a precooled insulated vessel of his own design. Dewar had invented this double-walled vessel in 1892. The vessel's insulating capacity came from the vacuum produced in the space between the inner and outer walls. Today the double-walled vacuum container used for liquid hydrogen and other low temperature fluids is called a dewar in honor of its inventor.[46] Mobile dewars, double-walled insulated tank trucks, were perfected by the Air Force in the 1950s.[47]

The Lewis group paid close attention to the evolving debate over the feasibility of liquid hydrogen as a rocket fuel. They visited Herrick Johnston, a national authority on liquid hydrogen, at Ohio State University. While a student at the University of California at Berkeley, he had contributed to the Harold Urey's discovery of deuterium or heavy hydrogen. In the 1930s, Johnston had built a hydrogen liquefier, or cryostat, at Ohio State. During World War II, Johnston's research contributed to the development of the atom and hydrogen bombs. At this time, Hsue-Shen Tsien, one of Professor von Kármán's students at Cal Tech, had proposed liquid hydrogen as a working fluid for a nuclear rocket.[48]

In the fall of 1950, the Lewis Laboratory sponsored a Propellant Selection Conference. Following the conference, the NACA authorized a subcommittee on rocket engines within the Power Plants Committee, chaired by Maurice Zucrow of Purdue University. Zucrow was frequently called upon for his advice on R & D for guided missiles. Zucrow invited Branch Chief Walter Olson to serve on the Research and Development Board for the Guided Missiles Committee for the Department of Defense. There Olson learned of

Meeting of the Navy Bureau of Aeronautics Panel on Propulsion and Fuels for Guided Missiles at the Rocket Test Station, Dover, New Jersey, 19 January 1951. Walter T. Olson, Combustion Branch Chief at Lewis (center), stands next to Maurice Zucrow of Purdue University (front row, second from left). Rocket Section head John Sloop is in back row on left.
USN NTI-3-1067

secret projects being conducted by General Electric and the Rocketdyne Division of North American Aviation. He returned to Lewis after these meetings determined to support more vigorously the work on high-energy propellants. Nevertheless, it was a tough sell at NACA Headquarters. Olson recalled that about 1955 Homer Newell of the Naval Research Laboratory encouraged the laboratory's participation in the Vanguard Satellite Program to commemorate the International Geophysical Year. Olson demurred, telling Newell that in view of the conservatism of headquarters, they were "lucky to be doing any rocket work at all."[49]

As historian Michael Gorn has pointed out in an essay to be published shortly by NASA, in the 1950s Hugh Dryden, the NACA's Director, was deeply engaged in reorienting the NACA to space-related research, despite a charter that limited it to advancing aeronautics. Between 1951 and 1957 the NACA's budget rose from $63 to $77 million. This small increase made it extremely difficult to fund space-related research, though initiatives were underway at all three laboratories. At Ames a new theory, verified by wind-tunnel experiments, found that blunt shapes dissipated heat. Actual rocket launches from the NACA's Pilotless Aircraft Research Station confirmed the superiority of the rounded nose cone for re-entry into the Earth's atmosphere. Langley was also deeply committed at this time to the development of the X-15, a hypersonic aircraft, capable of altitudes of 300,000 feet.[50]

One of the obstacles that Dryden faced was the unwieldy NACA committee structure which could only make recommendations, but carried no financial clout. Even though Zucrow, one of the country's foremost rocket experts, enthusiastically supported the laboratory's 1952 request for an appropriation for an $8.5 million large rocket engine facility to be built in a remote location in one of the western states, this request was turned down.[51] The proposed facility was to be used to test a 20,000-pound-thrust rocket using high-energy propellants, such as fluorine/hydrazine and fluorine/ammonia, and oxygen/hydrogen, as well as more traditional propellant combinations such as oxygen/gasoline and nitric acid/gasoline in engines up to 100,000 pounds of thrust.[52] In a memo to NACA Headquarters, Abe Silverstein noted Zucrow's "hearty endorsement of the planned new facility, emphasizing that it was exceedingly important that the NACA conduct its rocket research in engines of practical size as he noted the proposal provided for."[53] Zucrow's only reservation, it seems, was that 100,000-pound-thrust capability might not be large enough, since he envisioned engines of 200,000 to 300,000 pounds thrust within 10 years.

When these plans were scaled back to a modest $2.5 million, appropriated by Congress in September 1954, the NACA was left with the more experimental piece of the planned facility. The NACA's facility would be dedicated to testing high-energy propellants. Rather than a remote location in the far west, the proposed Rocket Engine Test Facility (RETF) was small enough to be built at the Lewis Flight Propulsion Research Laboratory. Construction began in May 1955 and was completed in 1957. Involved in the redesign were B. G. Gulick, D. W. Berg, D. A. Friedes, L. R. Marcus, J. H. Nitchman, O. J. Haas, O. J. Luchini, Dr. L. Gibbons, T. Reynolds, P. M. Ordin, S. Deutsch, and L. H. Rieman.[54]

As plans for the new facility evolved, Lewis Laboratory became more aggressive in its effort to become the nation's specialists in liquid hydrogen rocket technology. Silverstein allowed the group to purchase a small hydrogen liquefier from Arthur D. Little, an important engineering consulting firm in Cambridge, Massachusetts. William Rowe was sent to Cambridge for instruction from the firm on how to operate this new equipment, which liquefied hydrogen at a rate of up to 25 liters an hour. Still unable to produce liquid hydrogen in sufficient quantities to satisfy their research requirements, a few years later they jumped at the chance to purchase equipment left over from testing a "wet" hydrogen bomb in the South Pacific. This excess government equipment, owned by the Atomic Energy Commission, was stored at Edwards Air Force Base in California. Because of Rowe's previous experience with the Arthur D. Little liquefier, Sloop sent him out to Edwards. Let loose in the warehouse over one weekend, he tagged two rail carloads of equipment, instrumentation, valves, and plumbing for making liquid hydrogen. Rowe, and later Glenn Hennings, worked on setting up a system that could supply about 100 liters of liquid hydrogen per hour. This remained the laboratory's main source of liquid hydrogen until a new plant in Painesville,

Ohio, began producing liquid hydrogen that could be trucked to the laboratory in specially designed mobile dewars.[55]

Theoretically, hydrogen mated with fluorine promised the highest performance of any rocket propellant because of the combination's high combustion temperature and low exhaust molecular weight. Thrust of a rocket engine depends on the exhaust velocity of the combustion gases. Exhaust velocity is proportional to the square root of combustion temperature divided by molecular weight of the exhaust. "When we got through calculating, the best propellant combination by far—head and shoulders above everything else—was hydrogen with fluorine," recalled researcher William Tomazic.[56] With the hydrogen/fluorine combination, both high combustion temperature and low molecular weight contributed to high performance. Another advantage of hydrogen/fluorine was that its very low density meant that less fuel was required to achieve maximum specific impulse. This held out the tantalizing possibility of rockets with lower structural weight and higher performance.[57]

John Sloop and Howard Douglass became fluorine's champions. Fluorine was attractive because it spontaneously reacted with substances to unlock their chemical energy and liberate heat. The more heat, converted into kinetic energy by expansion through the rocket nozzle, the higher the potential thrust.[58] Fluorine, however, was a volatile and dangerous reactant. Researchers compared fluorine to the Hollywood swashbuckler Errol Flynn because of its propensity to attack whatever lay in its path. Fluorine took the finish off cars parked in the parking lot and ate through metal containers. The highly toxic gas posed serious risks to the environment—not to mention endangering the lungs and skin of researchers who ventured into its path.

Besides its reactivity, another unattractive feature was its expense. An extremely rare chemical, it had to be purchased as a gas and then liquefied prior to a test. At that time there was no instrumentation to test for a fluorine leak. After an aborted test had released fluorine into the atmosphere, Eugene Krawczonek recalled that before entering the test area he took the precaution of wearing a gas mask and his heavy overcoat. He discovered that fluorine immediately reacted with the wool in his coat, coating it with a sticky substance.[59]

A 1953 memo prepared by Edward Rothenberg for his boss, Branch Chief John Sloop, described the group's approach to designing a system to remove toxic hydrogen fluoride gas from the exhaust of ammonia/fluorine engines. It was first tried in Cell 14, a test cell capable of testing a 1,000-pound liquid hydrogen/fluorine rocket. Combustion gases were exhausted into a duct 30 feet in length and 20 inches in diameter. After a caustic solution was sprayed into the duct, the gases were vented to the atmosphere through a 20-foot-high, 36-inch-wide stack. A sample of the vented gases showed they had been rendered harmless. The only problem noted was that scaled up to treat the exhaust of a 20,000-pound-thrust rocket, the "scrubber" would be monstrous.[60]

Before the construction of the RETF, Cell 22 (also called the High Energy Rocket Systems Stand) was the largest and most advanced facility for rocket tests. Cell 22 had a large scrubber and two parallel test stands capable of testing rockets of up to 5,000 pounds of thrust. Liquid hydrogen testing began in November 1954.[61]

Lewis engineers working with high-energy fuels had no models to follow in designing these facilities. They were constantly challenged to explain unexpected explosions of test engines. One accident stood out in the memory of Frank Kutina, later head of RETF operations. This explosion not only destroyed the test stand, but also blew out the windows in the control room building and the offices of the 10-foot-by-10-foot wind tunnel located nearby, resulting in unwonted attention from upper management.

Kutina recalled they were perplexed by the explosion because it involved hydrocarbon (RP-1) fuel—not hydrogen—and it occurred a full five minutes after engine shutdown. Investigation revealed that the unburned fuel in the scrubber had combined with the oxygen from the ambient air to cause the explosion. New precautions included filling the scrubber with carbon dioxide gas prior to each engine firing, a costly procedure that was also used in the early phases of hydrogen testing in the RETF. Later, researchers were able to find a more efficient and economical way to prevent the explosion of unburned fuel.[62]

Engineers were able to test a scaled-down model of the Vanguard engine, which burned JP-4 with liquid oxygen. The Vanguard engine had a heat transfer problem that resulted in a meltdown of the walls of the thrust chamber. Researchers experimented with adding silicone oil to the fuel which, when burned, formed a protective coating of silicone oxide on the chamber walls. Later they would test a full-scale Vanguard engine in the RETF—the only full-scale engine ever tested in the facility.[63]

The first tests in Cell 22 of a liquid hydrogen/fluorine rocket proved frustrating. Despite regenerative cooling, the injector tended to melt down.[64] They had better results with tests of a regeneratively cooled hydrogen/oxygen thrust chamber in September 1957. William Tomazic recalled his elation over the results of the liquid hydrogen tests: "We had some interesting experiences there, and we pretty well proved, yes, the stuff is great! Performance is great! If you design properly, it cools beautifully."[65]

FROM SUNTAN TO THE RL10

While the NACA was testing liquid hydrogen rocket engines in Cell 22 and engaged in the building of the RETF, the Air Force spearheaded a much larger effort to develop a secret reconnaissance plane fueled with liquid hydrogen. Though cancelled in 1958, the Air Force program left behind not only an infrastructure for producing, transporting, and storing liquid hydrogen, but also a successful turbojet engine that ran on liquid hydrogen. This air-breathing engine later became the basis for the design of the first liquid hydrogen rocket engine—the RL10.

Hydrogen (which burns readily at extremely low pressures) seemed promising for high-altitude spy planes. To convince Air Force

Frank Kutina, later head of RETF operations, monitors a test in the control room of Cell 22, March 1957. Cell 22 was used for testing liquid hydrogen/liquid fluorine engines. Its scrubber served as the prototype for the much larger scrubber of the RETF.
NASA C-44591

officials that a liquid hydrogen power plant was feasible, Abe Silverstein and Eldon Hall produced a classified report, published in 1955 as "Liquid Hydrogen as a Jet Fuel for High-Altitude Aircraft."[66]

Following the publication of the Silverstein-Hall report, the Air Force funded the Suntan Project. Suntan was so secret that even Silverstein was kept in the dark about the full extent of the program. While spending approximately $100 million on Suntan, the Air Force funded a $1 million experimental liquid hydrogen aircraft engine test program at Lewis called "Project Bee."[67] Apparently, the Air Force regarded the Lewis effort as a low-cost effort that might foil Soviet intelligence. "Somebody is going to get wind of liquid hydrogen engine development sooner or later," Suntan's Air Force Chief Norman Appold said, "so we were very much willing to support and do anything that Abe and his folks wanted to do, because it gave us this cover."[68] Spending for Suntan greatly exceeded the entire budget of Lewis, which was about $22 million. It also exceeded NACA appropriations for 1956 of about $73 million.[69]

In addition to $1 million for the project, the Air Force supplied a B-57B bomber and two Curtiss-Wright J-35 engines. It gave Silverstein one year to prove that the aircraft could be adapted to fly with liquid hydrogen fuel. The rocket group's experience with liquid hydrogen proved vital to the success of the project. Howard Childs headed the Guidance Panel charged with establishing the design criteria to be carried out by an Operations Group headed by Paul Ordin. Ordin, a 1940 graduate of City College of New York in chemical engineering, already had more than a decade of experience with liquid hydrogen.[70]

The B-57B bomber was modified so that one engine could burn liquid hydrogen fuel, code-named "X-35." During the first flight of the aircraft on 19 December 1956—one year from the date of the start of the project—the pilot switched from JP-4 fuel to liquid hydrogen twice, but the plane failed to maintain its speed. A second flight had only partly satisfactory results. During the third flight on 13 February 1957, the engine was successfully operated for 20 minutes on liquid hydrogen at an altitude of 49,500 feet and a speed of Mach 0.72.[71]

Meanwhile, a far more ambitious development program for the Air Force reconnaissance plane had made rapid progress. The Pratt & Whitney Aircraft Engine Company was selected by the Air Force to build a turbojet engine for Suntan. It completed a prototype in August 1957. At the same time, the Air Force financed the construction of new liquid hydrogen plants in Trenton, New Jersey (later cancelled), Painesville, Ohio, and Bakersfield, California. These plants increased the nation's production capacity from several hundred pounds to 1,500 pounds per day. The Air Force also supported construction of two plants adjacent to an extensive new Pratt & Whitney engine test center in West Palm Beach, Florida. These plants, known as "Mama Bear" and "Papa Bear," produced respectively 7,000 and 60,000 pounds of liquid hydrogen per day.[72] Mobile dewars, later used by the RETF for transporting liquid hydrogen, were also developed as part of the Suntan project.

Despite the project's cancellation in 1958, Suntan proved an important step toward the realization of a liquid hydrogen rocket. Shortly after the Soviet Union launched Sputnik, the Advanced Research Projects Agency (ARPA) granted a contract to General Dynamics for the development of Centaur, a liquid hydrogen upper-stage rocket. The RL10 engine, designed by Pratt & Whitney for the Centaur rocket, grew directly out of the turbojet engine it had designed for the Suntan Program. Commenting on the appropriateness of the choice of Pratt & Whitney for the Centaur engine project, Air Force Colonel Norman Appold said:

> But having developed all the capabilities to pump and handle liquid hydrogen, to meter hydrogen, understanding its combustion properties better than virtually anybody else in the country, and particularly in terms of the high level of classification in the Suntan Program, it seemed more practical and economical to encourage Pratt & Whitney to enter the rocket business than to take the technology and give it to someone like North American or Aerojet General.[73]

Pratt & Whitney thus entered the rocket business by the back door, but its experience with liquid hydrogen became an important national asset. The RL10 engine for Centaur would later prove the feasibility of liquid hydrogen as a fuel for upper-stage rockets.

For Lewis Laboratory, Project Bee represented an interesting detour by rocket researchers into the field of air-breathing engines. It had demonstrated that an aircraft could be flown with liquid hydrogen fuel; it also revealed that further research was needed in the areas of liquid hydrogen storage, insulation, instrumentation, and pumping. In assessing the flight research program, Olson reflected that it pushed rocket researchers to approach liquid hydrogen technology as a system. "We had to have a whole lot of things that all worked: tank, line, control, pump, combustion, the whole ball of wax. I think much of the laboratory work up until then had been piecemeal on components and processes. I would say that a project like that began to demonstrate the strength of bringing these things together in a working system."[74] The experience of managing Project Bee would prove invaluable to the Lewis staff when they took over management of the NASA's Centaur Program in 1962. Meanwhile, the staff would focus on getting the new Rocket Engine Test Facility up and running.

NOTES

1. John Sloop, "High Energy Rocket Propellant Research at the NACA/NASA Lewis Research Center, 1945-60: A Memoir," AAS 89-226 in AAS History Series, 1989, vol. 8, 273.

2. John Sloop, "NACA High Energy Rocket Propellant Research in the Fifties," AIAA 8th Annual Meeting, Washington, DC, 28 Oct. 1971 [typescript], 2-3.

3. Reprinted in Robert Goddard, *Rockets* (American Rocket Society, lst ed., 1946; AIAA, facsimile ed., 2002).

4. On early rocket societies, see Frank H. Winter, *Prelude to the Space Age: The Rocket Societies 1924–1940* (Washington, DC: Smithsonian Institution Press, 1983).

5. For an overview of early rocket development, see Tom D. Crouch, *Aiming for the Stars: Dreamers and Doers of the Space Age* (Washington, DC: Smithsonian Institution Press, 1999).

6. Winter, *Prelude to the Space Age,* 84.

7. Ibid., 99-102. See also John Tascher, "U.S. Rocket Society Number Two: The Story of the Cleveland Rocket Society," *Technology and Culture 7* (1966): 48-63.

8. Frank Malina, "The U. S. Army Air Corps Jet Propulsion Research Project, GALCIT Project No. 1, 1939–1946: A Memoir," *History of Rocketry and Astronautics,* R. Cargill Hall, ed., AAS History Ser., vol. 7, part 2, 1986, 159.

9. See Willy Ley, *Rockets, Missiles, & Space Travel* (New York: The Viking Press, 1958), 61-63.

10. On Malina and JPL, see Clayton Koppes, *The JPL and the American Space Program* (New Haven: Yale University Press, 1982). Because of science fiction connotations surrounding the word "rocket," the group referred to rockets as "jets." After the war, Malina's group founded the Jet Propulsion Laboratory (JPL) on the site of its former rocket tests.

11. Vannevar Bush, *Modern Arms and Free Men* (NY: Simon and Schuster, 1949), 85.

12. Walter Olson interviews at Greater Cleveland Growth Assn., 17 July 1984, and 29 October 1984, Walter T. Olson Bio. file, NASA Historical Reference Collection, Washington, DC.

13. Frank J. Malina and Martin Summerfield, "The Problem of Escape from the Earth by Rocket," *Journal of the Aeronautical Sciences* 14 (Aug. 1947). See also comment by Theodore von Kármán, "Martin Summerfield: Ten Years as Editor," *American Rocket Society Journal,* 31 (1961): 1182.

14. See James R. Hansen, *Engineer in Charge: A History of Langley Aeronautical Laboratory* (Washington, DC: NASA Special Publication-4305, 1987), 63.

15. R .O. Miller and P. W. Ordin, "Theoretical Performance of Rocket Propellants Containing Hydrogen, Nitrogen, and Oxygen," NACA RM E8A30, May 1948.

16. Sloop, *Liquid Hydrogen,* 74; V. N. Huff, S. Gordon, and V. E. Morrell, "General Method and Thermodynamic Tables for Computation of Chemical Reactions," NACA Report 1037, 1951; supercedes V. N. Huff and V. E. Morrell, "General Method for Computation of Equilibrium Composition and Temperature of Chemical Reactions," NACA TN 2113, June 1950.

17. Sloop, "High Energy Rocket Propellant Research," 274.

18. Ibid.

19. Virginia Morrell interview, Middleburg Heights, OH, 6 Dec. 2002.

20. Martin Summerfield, "The Liquid Propellant Rocket Engine," *High Speed Aerodynamics and Jet Propulsion* (Princeton: Princeton University Press, 1959), vol. 12, 518.

21. Walter T. Olson Interview by John Sloop, p. 29, cited above, chap. 1, note 3.

22. See *Exploring the Unknown: Selected Documents in the History of the U.S. Civil Space Program,* vol. 1 (NASA SP-4218, 1995), 67-68, reprinted from A. A. Blagonravov, Editor in Chief, *Collected Works of K. E. Tsiolkovskiy, Volume II – Reactive Flying Machine.*

23. Sloop, *Liquid Hydrogen,* 269.

24. Ibid, 31-58.

25. Ibid., 28.

26. Ibid., 75.

27. Taped telephone interview with William Rowe, 10 Dec. 2002.

28. Ibid.

29. Vearl N. Huff, Clyde S. Calvert, and Virginia C. Erdmann, "Theoretical Performance of Diborane as a Rocket Fuel," NACA RM E8117a, Jan. 1949; William H. Rowe, Paul M. Ordin, and John M. Diehl, "Investigation of Diborane-Hydrogen Peroxide Propellant Combination," NACA RM E7K07, 1948; Rowe, Ordin, and Diehl, "Experimental Performance of Diborane-Oxygen in a 100-Pound-Thrust Rocket Engine," NACA RM E9C11, May 1949.

30. Bureau of Aeronautics to Director of Aeronautical Research, memos for 2 June 1948 and 7 March 1949, NACA Programs/Contingency Plans, 1951–58, NASA Glenn Records, Box 295, RG 255, National Archives and Records Administration, Chicago, Illinois. Hereafter cited as NACA Programs/Contingency Plans. Project Research Authorizations between 1947–1957 are in Box 502 of the same collection.

31. "Combustion Research Program, NACA Lewis Laboratory active as of September 1949," NACA Programs/Contingency Plans, 1951–58.

32. Ibid. See also, Sloop, *Liquid Hydrogen,* 75.

33. R. J. Minty to Walter Olson, "Rocket Propellant Combinations," 23 May 1949, NACA Programs/Contingency Plans, 1951–58.

34. "Combustion Research Program NACA Lewis Laboratory active as of September 1949," NACA Programs/Contingency Plans, 1951–58.

35. Ibid.

36. Theodore von Kármán, *The Wind and Beyond* (Boston: Little Brown & Company, 1967), 252.

37. See "Transmittal of Rocket Research Program for possible distribution to members of the NACA Special Subcommittee on Rocket Engines," prepared by John L. Sloop, 23 Sept. 1954, Rocket Engine Research Program, 1952–1954, NACA Programs/Contingency Plans, 1951–58.

38. Marcus F. Heidmann and Jack C. Humphrey, "Fluctuations in a Spray Formed by Two Impinging Jets," *Journal of the American Rocket Society* 22 (1952): 127-167. Controversy over the paper is indicated in the transcript of comments that followed the paper.

39. Theodore Male, William Kerslake, and Adelbert O. Tischler, "Photographic Study of Rotary Screaming Oscillations in a Rocket Engine," NACA RM E54A29, 1954.

40. "Review of Rocket Research at Lewis," 9 Feb. 1954, NACA Programs/Contingency Plans, 1951–58.

41. Ibid.

42. Krafft Ehricke interview by John Sloop, 26 April 1974. Ehricke Bio. file, NASA Historical Reference Collection, Washington, DC.

43. K. A. Ehricke, "Comparison of Propellants and Working Fluids for Rocket Propulsion," *American Rocket Society* 23 (1953): 287-96, 300.

44. Ehricke interview.

45. Sloop, *Liquid Hydrogen,* 251.

46. Sloop, *Liquid Hydrogen,* 251, 266-67. Also, "James Dewar," in Charles Coulston Gillispie, ed., *Dictionary of Scientific Biography* (New York: Charles Scribner's Sons, 1981), 78-81.

47. Sloop, *Liquid Hydrogen,* 251.

48. On Johnston, see Sloop, *Liquid Hydrogen,* 13-29. See also H. L. Coplen, "Large-Scale Production and Handling of Liquid Hydrogen," *Journal of the American Rocket Society* 22 (Nov.–Dec. 1952): 309-322, 338; James Dewar, "Project Rover: A Study of the Nuclear Rocket Development Program, 1953–1963," Ph.D. Dissertation, University of Kansas, 1974; and Richard Rhodes, *Dark Sun: The Making of the Hydrogen Bomb* (New York: Simon & Schuster, 1995), 488.

49. Sloop interview with Olson, 43.

50. Michael Gorn, "Who Was Hugh Dryden, and Why Should We Care?" in Virginia Dawson and Mark Bowles, eds., *Realizing the Dream of Flight* (Washington, DC: NASA Special Publication-2005-4112, impress).

51. Sloop, *Liquid Hydrogen,* 80.

52. John Sloop, "Specifications and suggestions for proposed rocket test facility," 14 April 1952, Joe Morgan Papers, Box 14, History, Folder 1 of 2, RETF Record Group, NASA Glenn Research Center, Cleveland, Ohio, hereafter cited as GRC.

53. Memo from Silverstein to Rothrock, "Discussion with Dr. M. J. Zucrow of proposed addition to rocket research facilities," 24 Nov. 1952, NACA Programs/Contingency Plans, 1951–58.

54. "Meeting to Discuss S-40 Rocket Lab Facilities," memo for record, 10 Sept. 1954, Joe Morgan Papers, Box 14, folder 2, RETF Record Group, GRC. See also *Wing Tips* (28 Oct. 1955). Rocketeers listed included H. Douglass, R. Priem, M. Heidmann, F. Salzano, H. Price, M. Lieberstein, C. Auble, W. Tomazic, D. Nored, J. Sloop, I. Pass, G. Morrell, C. Feiler, G. Kinney, J. Rollbuhler, A. Tischler, E. Krawczonek, L. Baker, J. Bahan, E. Rothenberg, C. Bibbo, F. Kutina, D. Ladanyi.

55. Rowe interview, 10 Dec. 2002.

56. Tomazic interview, 23 Feb. 2003.

57. See comparative discussion by A. A. McCool and Keith B. Chandler, "Development Trends of Liquid Propellant Engines," in *From Peenemünde to Outer Space* (Huntsville, AL: The George C. Marshall Space Flight Center, 1962), 289-307.

58. John Sloop, "Fluorine—the Oxidizer with a Potential in Space Propulsion," speech delivered at the opening of General Chemical's Fluorine Plant, Metropolis, IL, 30 Oct. 1958, John L. Sloop (Propulsion in the 1950s), misc. papers, NASA Historical Collection, Washington, DC.

59. Eugene Krawczonek interview, 23 Jan. 2003.

60. [Paul M. Ordin], "Memorandum for Chief, "Hydrogen fluoride scrubber development for fluorine-burning rocket Engines," 13 Nov. 1953, Box 1, file 40, Scrubber memos, RETF Record Group, GRC.

61. Sloop, *Liquid Hydrogen,* 198. On the role of Lewis Research Center in developing liquid hydrogen technology, see Roger E. Bilstein, *Stages to Saturn, A Technological History of the Apollo/Saturn Launch Vehicles* (Washington, DC: NASA Special Publication-4206, 1980), 129-137.

62. Verbal comments (not taped) on manuscript by Frank Kutina, 9 April 2004. Kutina remained lead operations engineer for RETF until 1966, when he was transferred to the Aeronautics Division to head a review of aeronautical test facilities.

63. Frank Kutina interview, 26 March 2004.

64. Sloop, *Liquid Hydrogen,* 88.

65. William Tomazic interview, 23 Jan. 2003.

66. NACA RM E55 C28a, 15 April 1955. See also Sloop, *Liquid Hydrogen,* 98-102.

67. Colonel Norman C. Appold interview by John Sloop, 4 Jan. 1974, 79, N. C. Appold Bio. file, NASA Historical Collection, Washington, DC.

68. Ibid.

69. Alex Roland, *Model Research: The National Advisory Committee for Aeronautics* (Washington, D.C.: NASA Special Publication-4103) vol. 2, 475.

70. Paul Ordin interview by John Sloop, 30 May 1974, misc. biographical interviews, K-O, NASA Historical Reference Collection, Washington, DC.

71. Ordin interview.

72. Project Suntan, "General Discussion by Col. N. C. Appold, Lt. Col. J. D. Seaberg, Maj. A. J. Gardner, Capt. J. R. Brill," transcribed Feb. 1959, 15, Appold, N.C. Bio. file, NASA Historical Collection, Washington, DC.

73. Appold interview, 56-57.

74. Olson interview, 35-36.

chapter 2

RETF: Complex, Versatile, Unique

AFTER SPUTNIK ROCKET RESEARCH at Lewis took center stage. More than 300 engineers from across the country flocked to a classified NACA Flight Propulsion Conference held at the laboratory in November. The Lewis rocket staff presented papers on various aspects of rocket design, including propellants, cooling, and turbopumps. John Sloop, A. S. Boksenbom, Sanford Gordon, Robert Graham, Paul Ordin, and Adelbert O. Tischler discussed propulsion requirements for Earth satellites and even suggested the possibility of landing on the Moon. Frank Rom, Eldon Sams, and Robert Hyland broached the idea of developing nuclear rockets, while W. C. Moeckel, L. V. Baldwin, Robert English, Bernard Lubarsky, and S. H. Maslen focused on satellites and space propulsion systems.[1]

The most dramatic presentation, however, concerned the extraordinary potential of liquid fluorine as an oxidizer with liquid hydrogen. Few in the audience knew that right up to the night before the presentation, the paper's authors, Howard Douglass, Harold Price, and Glenn Hennings, had not yet been able to confirm theoretical predictions in experimental hardware. Very early on the morning of the presentation, unknown to Douglass, who was delivering the paper, a rocket test in Cell 22 finally succeeded. Just before Douglass was called to the podium, Price handed him a note and grabbed one of Douglass's slides and inserted a key data point with a grease pen. Douglass then nonchalantly informed the audience of the first successful test of a regeneratively cooled hydrogen/fluorine rocket. This created a flurry of excitement.[2]

Douglass and his boss John Sloop saw great potential for a liquid hydrogen/liquid fluorine rocket. They no doubt hoped that they would soon be able to test an engine fueled with the combination in the new Rocket Engine Test Facility whose systems were still being checked out. Like many facilities built during the NACA era, the Rocket Engine Test Facility was designed in-house. It evolved from ideas and design concepts tried out in the laboratory's smaller test cells. The RETF's most immediate model was Cell 22.

NUTS AND BOLTS OF THE RETF

The RETF consisted of a cluster of buildings and propellant storage and handling facilities that covered about 10 acres. The test cell, often referred to colloquially as the "South Forty," or "S-40," was located in the center of a 40-acre parcel of land south of the central area of the laboratory. Ground was broken in the fall of 1955. Project engineer George Kinney, assisted by William Anderson and Lou Rieman, supervised construction by the H. K. Ferguson Company. The RETF cost $2.5 million and took two years to build.

The RETF enabled research engineers to experiment with innovative designs and concepts, analyze successes and failures, and generate data cost effectively by testing components and sub-scale engines. Because it was a research facility, the RETF differed from other military and industrial rocket engine test facilities which were used exclusively for missile development. For example, in 1957 the Army Ballistic Missile Agency in Huntsville, Alabama, completed a large engine test stand called the "Static Test Tower." This test stand was capable of testing the 75,000-pound Redstone missile.

Typically, RETF test engines had 4.8-inch chambers with 2.62-inch throats, or 10-inch chambers with 7.6-inch throats. By feeding the propellants into the engine under high pressure, an engine with a 4.8-inch throat, for example, could produce thrusts from 17,000 pounds up to 20,000 pounds. Later this was increased to 50,000

Engineer Walter Russell (left) from the Fabrication Division inspects a regeneratively-cooled thrust chamber for a 20,000-pound-thrust rocket engine while a technician cleans the face plate of an injector in the shop of Building 100, 15 Dec. 1958.
NASA C-49343.

pounds. Even by the standards of the mid-1950s, this was not a large thrust capability. What made the facility unique was its ability to handle a wide range of high-energy propellants. Fuels appropriate for testing in the new facility included hydrogen, ammonia, and ammonia/hydrazine mixtures, gasoline and other hydrocarbons, and alcohol. Oxidants included liquid oxygen, liquid ozone, mixtures of oxygen and ozone, liquid fluorine, and mixtures of oxygen and fluorine.

Despite, or perhaps because of its relatively small size, the RETF proved an extremely versatile and economical research facility, especially suited for experimental investigations. The test cell was located on a natural plateau on the edge of a ravine. Because the test area was below grade, gravity could be used to assist in handling the large quantities of water needed for removing contaminants from rocket exhaust. Designers placed the fuel pits for liquid oxygen and JP-4 (also referred to as RP-1) adjacent to the test stand. However, because of concern over the volatility of hydrogen, the hydrogen run tank was originally sited above ground outside the test cell. Although this location made it easy to fill the tank from a mobile dewar, which could drive up onto the test cell apron, it made the tank more vulnerable than if it had been placed underground. Engineers learned this lesson the hard way. When an early test engine blew up, a piece from the wall of the test cell severed one of the tank feed lines. This resulted in the loss of the entire supply of liquid hydrogen.

Transported to the laboratory in mobile or "roadable" dewars, propellants were pumped into storage dewars in the upper area. Just before a test, they were transferred to run tanks in the lower, or test stand area. The run tanks were then pressurized with a high-pressure gas (hydrogen or helium) that would force the propellants into the rocket engine injector for combustion. Nitrogen, because of its inert properties, was used for operation of valve actuators, purging of electrical boxes and valve bodies, purging of fuel lines and also for purging of the rocket engine at the end of each firing cycle. Liquid nitrogen was pressurized and then forced through a network of pipes to the vaporizer. There it gradually warmed until it boiled, forming high-pressure nitrogen gas. This inert, non-flammable gas was then piped throughout the RETF.[3]

During a test run, fuel and oxidizer supply were forced into the test engine at a controlled rate. Cryogenic liquid oxidizers required special precautions. Their stainless-steel tanks, located in the oxidant propellant pit, were suspended within insulated tanks that were automatically kept full of liquid nitrogen during and after they were loaded. This liquid nitrogen "bath" kept them cold enough to prevent vaporizing.

The walls and roof of the test cell were covered with corrugated cement asbestos panels. Originally, the panels were lightly fastened to steel I-beams, which could be replaced if an explosion blew them out. However, after the panels were blown off twice because of hydrogen escaping into the test cell, it was decided to leave the sides

After a liquid hydrogen explosion in early 1958, the test stand had to be completely rebuilt.
NASA C-4937

partially open during a test to prevent the dangerous buildup of unburned propellants.[4] In May 1959, the facility was modified to allow the walls of the test cell to be rolled up during testing. During inclement weather, the sides were rolled down to provide shelter when preparing hardware for testing.[5] During and between tests, the cell doors were kept open on the west, south, and upper east walls to prevent buildup of fuel vapors that might explode. The east door and fuel pit louvers in the upper north wall were also kept open. An exhaust fan provided additional ventilation.

A small room, called the instrument or terminal room (where all the instrumentation and control wires from the rocket engine test stand and other facility functions terminated) was located adjacent to the test cell. Two observers were generally stationed there during a test. To protect them, the room was pressurized to prevent dangerous fuel vapors from leaking in. It had one and one-half foot concrete reinforced walls and was equipped with a unique explosion-proof observation mirror.

From the terminal room, amplifiers sent the data over wires to the control room in the rocket operations building. The one-story building, referred to as Building 100, was located 1,700 feet from the test cell. It had a spacious control room, instrument room, shop for fabricating hardware, offices for about 32 people, and conference and drafting rooms. Later, raw data was transmitted to computers located in the 10-foot-by-10-foot supersonic wind tunnel office building (Building 86).

A blockhouse located 100 yards to the north of the cell served as an observation post for an additional technician. A television camera mounted on the blockhouse in the 1970s made it possible to eliminate this hazardous assignment. Support structures added in the 1960s included a propellant transfer and storage facility and a building to house a cryogenic vaporizer and compressor.

The facility's enormous scrubber/silencer was the RETF's most unique feature. Its design evolved from years of experience testing fluorine as an oxidizer with liquid hydrogen in Cell 22.[6] The RETF scrubber consisted of a large horizontal tank containing seven water spray bars. The 77-foot-long scrubber tank was connected to a vertical exhaust stack 20 feet wide at its base and 58 feet high. By the mid-1960s, it was necessary to add stack height when test programs were expanded to include fuels with more environmentally hazardous exhaust products, such as nitrogen tetroxide (N_2O_4) and unsymmetrical-dimethylhydrazine (UDMH). At that time, the stack was tapered

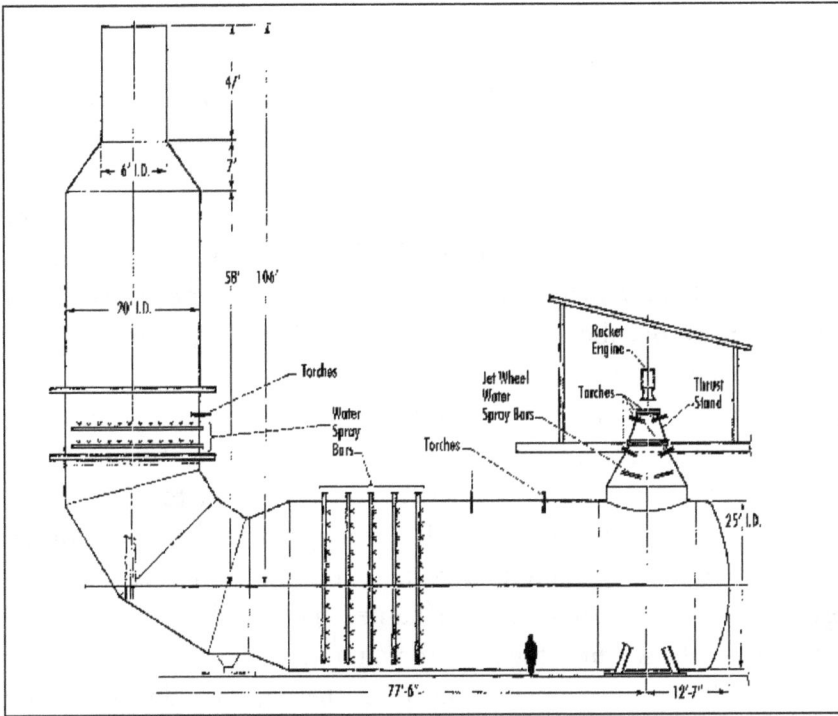

Cross-section of exhaust duct, or scrubber. Five spray bars were located in the horizontal section, with two additional bars in the vertical section. Note the location of the jet wheel in the exhaust duct under the stand and the seven hydrogen torches, used to burn off excess fuel in the scrubber.
NASA TN D-3373, 1966.

to a width of 6 feet, reaching a total height of 106 feet. The added height and greater velocity of the exhaust products leaving the stack kept contamination of the atmosphere within safe limits.[7] During testing, the spray bars in the tank delivered 50,000 gallons of water per minute for cooling the rocket exhaust.

Another unique feature of the RETF was a device called a "jet wheel," a scheme for injecting water directly into the core of the high-temperature, high-velocity rocket engine exhaust in order to reduce the temperature. Again, the jet wheel evolved from the determination of the facility's designers to solve a problem unique to experimentation with high-energy fuels. Water was injected into the exhaust stream at a point immediately below the engine through an array of pipes, arranged like the spokes of a wagon wheel. The pipes were angled at approximately 45 degrees downstream relative to the rocket exhaust. The round pipes were flattened at their ends to form a crude nozzle to reduce the aerodynamic loads. They terminated close to, but downstream of the rocket exhaust nozzle.

The jet wheel allowed the exhaust to be handled by the ordinary carbon steel ducting of the RETF. The hot gases, emerging from the rocket nozzle at velocities of 9,000 to 12,000 feet per second and temperatures of about 6,000 degrees Fahrenheit, were quickly cooled to steam temperature and slowed to a velocity of 25 feet per second. Water from the spray bars condensed the steam, allowing only non-condensable exhaust gases to emerge from the top of the stack. Excess water from the scrubber was ducted to a detention tank, where poisonous chemicals, principally hydrogen fluoride, were treated with calcium hydroxide. The residue, inert calcium fluoride precipitate, could then be hauled away, while the cleansed water was released into the ground water system.[8]

Controls and instrumentation for gathering data during testing were essential to the operation of the facility. The facility's designers not only adapted instrumentation used in the laboratory's wind tunnels, but also were forced to develop unique instrumentation as the project evolved. The most important readings were thrust, fluid flow, pressures, and valve positions. Control of the flow of oxidant and fuel was essential in order to produce the desired ratio of flows and combustion pressure in a rocket engine.

In designing the smaller test cells, Lewis researchers had found that manual control of propellant flows and chamber pressure during rocket tests was difficult. The facility's designers realized that in the new scaled-up facility, failure to maintain test parameters would drive up the expense of tests, since one second of operation with fluorine cost more than $800. Because of the difficulty and urgency of this problem, the design team requested research funds to develop an automatic system, since there was little commercially available equipment that suited their needs.[9]

RETF's unique control system consisted of precise, timed, and sequenced controls (required during the actual firing of the rocket), auxiliary controls (not actually operated during firing), safety controls, facility controls, and safety alarm equipment. The operations building was connected to the terminal room and to the test cell by a system of duct banks through which electrical signal cables were routed. If a problem were suspected, signals transmitted to the test cell via this system could terminate the test sequence at any point during a test.

In contrast to the location of some of the nation's other test facilities in remote areas of New Mexico, California, Nevada, and Florida, the RETF was built in a relatively populated area. In addition to minimizing emissions, engineers tackled with considerable gusto the problem of silencing the rocket tests. Prior to construction of the RETF, Lewis engineer T. W. Reynolds made a complete analysis of the noise problem in a memo dated 25 April 1954. His analysis cited a report of an independent contractor who estimated that the sound from a 20,000-pound thrust-engine would be about 190 decibels. However, attenuation of the sound by the scrubber would reduce it to a maximum sound intensity of approximately 105 decibels. This was considered a tolerable level.[10]

Safety features included several warning systems. For example, green, yellow, and red lights in the test area provided visual indications of various levels of alert. Barriers were put up before tests. Thirty seconds before actually firing the rocket engine, a siren would go off and continue until the end of the test. To protect the test stand, a gravity-fed water deluge could be activated. Finally, a public address system could be used to warn workers in various propellant storage areas, at the test stand, and in the operations building.

The test cell and the areas where hydrogen was stored were monitored by a 10-station combustible alarm sniffer system, made commercially for monitoring the air quality in mines. During a test, sniffers continuously sampled for hydrogen gas at five stations: under the engine, at the jet wheel, near the duct purge valve, near the elbow, and at the top of the 20-foot vertical section. Four addition-

SECTIONAL VIEW OF HIGH-ENERGY ROCKET TEST FACILITY

Cross-section showing test stand, scrubber and oxidant pit (left). The scrubber cleansed the exhaust gases and helped to silence rocket engine noise.

al sniffer stations were installed in the mid-1960s to make the system even more effective.

Gauges for the sniffer system were located in the shop as well as in the control room of Building 100. An audible alarm was sounded in the cell when the hydrogen content of the air in the cell went above one percent of the lower explosive limit. At the same time, a red light appeared on the control room panel and a sniffer system panel in the shop. If a leak were detected, the test run would be delayed until the source was found and repaired.

Early testing revealed another unanticipated problem. The large liquid hydrogen tank was typically pressurized to over 1000 psig prior to and during a test. At the end of the test, the tank needed to

A technician looks from the test cell doorway toward the scrubber, 1957.
NASA C-45734.

be depressurized. A large valve was opened, liberating a large quantity of hydrogen gas, initially at high velocities. This hydrogen gas mixed readily with air. Occasionally, static electricity would cause the gas mixture to ignite, resulting in 20-foot-diameter fireballs. The solution to this problem was to add a small bleed valve to allow the tank to be vented slowly. Fireballs became a thing of the past.[11]

Test engineers aimed for peak performance from a rocket engine during a test, a requirement that mandated running the engine with excess fuel. During the firing of an actual rocket system, exhaust from the rocket nozzle dissipated into the atmosphere. However, in the RETF there was the danger that fuel-rich exhaust, funneled into the scrubber, could explode. To make sure there was no oxygen left in the scrubber to allow combustion, the scrubber was flooded with 12 tons of carbon dioxide before a test and sometimes during a test. The carbon dioxide was blown out of the duct at its conclusion. At a cost of $128 per ton, this drove up the cost of testing. Another drawback was that it took about an hour to pump the carbon dioxide into the exhaust duct prior to a test. "It didn't take us that long to do a cursory scan of the data and decide what we wanted to do next," recalled lead operations engineer Frank Kutina. A graduate of Case Institute of Technology, Kutina had worked in the rocket area since 1954 and knew that the problem of treating the unburned fuel in the scrubber needed to be solved if the RETF were to become cost effective.

Kutina proposed the installation of seven small torches in the scrubber to provide a continuous source of ignition to burn off the excess fuel. "We did some small test work with just a pipe. We proved the concept to ourselves and then we went before the Safety Committee and sold them on the idea." This important innovation eliminated the need for carbon dioxide purges. Two torches were mounted below the engine at opposite ends of the test stand, two above the jet wheel, two on the top of the horizontal section, and one above the seventh spray valve in the 20-foot vertical section. The torches, lit by spark coils and spark plugs, were instrumented with two thermocouples welded to their tips. These thermocouples sent data to the torch control panel in the control room to assure proper operation and provide a safety interlock. The torch valves were remotely operated and constantly monitored before, during, and after a test run. Since the hydrogen torches could be turned on and off at will, and only small amounts of hydrogen were required, the cost in terms of hydrogen fuel was negligible. The new system greatly increased the productivity of the facility and dramatically reduced costs.[12]

The following passage from a safety procedure memorandum written in 1964 by operations engineer Larry Leopold and carefully preserved for 40 years by operations engineer Joe Morgan conveys a sense of the complexity of operation and technical virtuosity that running the facility demanded:

Our firings of rocket engines always dump free H_2 into the scrubber. In addition, we open our H_2 valves about 2 seconds before the LOX [liquid oxygen] fire valve and ignite the H_2 with a minimum of flow of F_2, to obtain a low flow lazy H_2–rich flame, which dumps free H_2 into the scrubber. The duct purge fan is off at this point, and the duct purge valve is shut. The oxygen available in the scrubber amounts to about 1150 lbs. spread throughout the entire scrubber. . . . After a run, the MSA [Mine Safety Appliance, or sniffer] stations near the engine generally read 40 to 60% L.E.L. [Lower Explosive Limit], which rapidly drop to zero a few minutes after as the H_2 is consumed by the torches. This is a smooth, quiet operation—obviously not a detonation. Then the duct purge valve and fan are turned on to purge fresh air through the scrubber.[13]

A commercial oxygen analyzer, made by the Beckman-Pauling Company, continuously sampled the oxygen level in the scrubber, sending data that was recorded on strip charts in the control room. Safety was always of paramount concern, but dealing with fluorine required special precautions. During tests involving fluorine, the back gate was also informed in case emergency vehicles were needed. Fluorine was stored in a special fluorine trailer parking area located east of the water reservoir at the south end of the upper road. In case of a spill, a water spray system would inert the fluorine (F_2), converting most of it to hydrofluoric acid that was neutralized in a limestone pit under the gravel of the parking area. During a typical test of liquid hydrogen/liquid fluorine propellants, sprays in the scrubber diluted the fluorine with about 27,000 pounds of water per run, yielding about a half pound per minute of fluoride in the detention tank water.[14] While experiments that involved fluorine as an oxidizer required no ignition source, engineers also discovered they could use a small amount of fluorine for ignition of other propellant combinations. The flow rate for the fluorine igniter was 0.6 pounds per second. This ignition system was used until 1973, when the RETF was converted to electrical ignition.

FIRST TESTS

As lead operations engineer for RETF, Frank Kutina was responsible for supervising the final phases of construction and its "shakedown." The shakedown involved testing the various subsystems leading up to the first test, which occurred on 15 August 1957. The first test involved firing a 20,000-pound thrust rocket engine with

A technician checks the installation of the 20,000-pound-thrust rocket engine on the test stand, October 1957.
NASA C-45869.

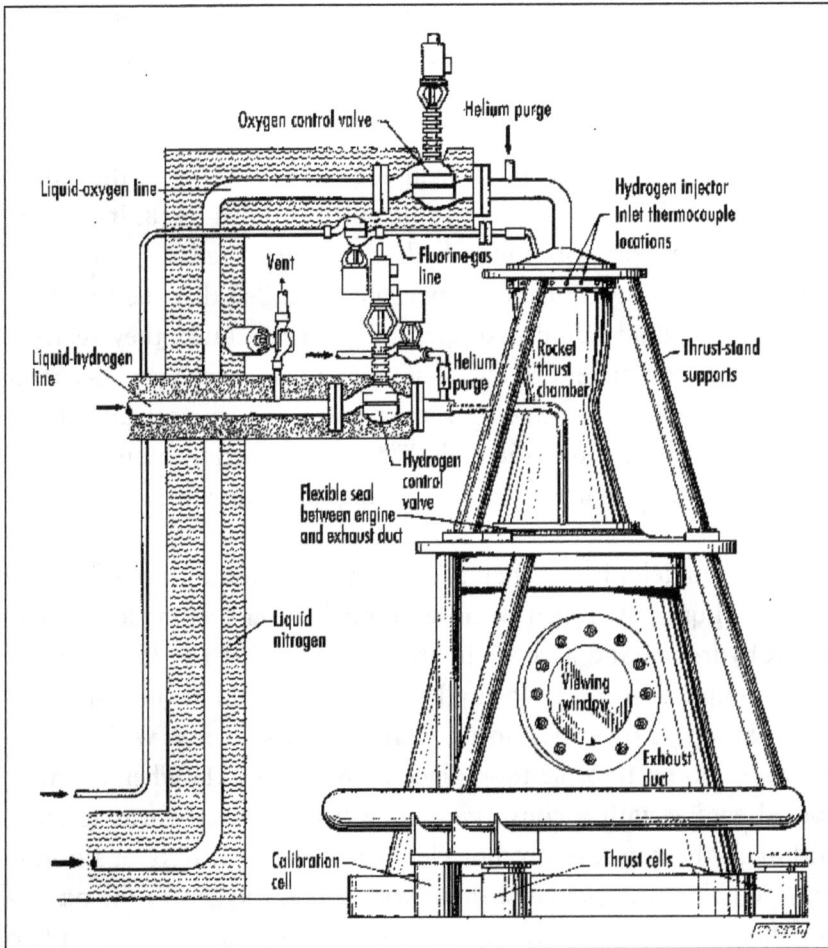

Diagram with rocket thrust chamber mounted on the test stand. Liquid hydrogen and liquid oxygen feed lines can be seen (left). Rocket test stand legs have been split to facilitate easier access to hardware.
NASA TM X-253, 1960.

kerosene-based JP-4 and liquid oxygen. Kutina noted in the run log that the test "went very smooth."[15] Though no doubt pleased with the results of this test, the facility's operators were after bigger quarry. How would the facility handle more exotic fuels like gaseous and liquid hydrogen, and especially liquid fluorine? While the RETF staff awaited delivery of the liquid hydrogen run tank, on 14 November they tested a water-cooled rocket engine fueled with gaseous hydrogen and liquid oxygen. On this test, the propane torch did not light, and the cooling tubes leaked water. When tried five days later, the engine blew up.[16]

From the notations in the run log, it appears that the liquid hydrogen tank was delivered and installed in April 1958. It is noted that 4,000 gallons of liquid hydrogen were used that month.[17] Operations engineer Eugene Krawczonek recalled that the first time they tried to load liquid hydrogen into the run tank, they realized they had no instrument to tell how much liquid hydrogen was actually in the tank. They came up with the idea of sticking a tube down the opening in the top of the tank. When it touched the liquid hydrogen, it "shot up a puff of smoke," letting them know the level of the fuel.[18]

Unfortunately, it is difficult to fix the exact date of the first test run with liquid hydrogen because no distinction was made between liquid hydrogen and gaseous hydrogen in the first run log. A document compiled in the 1970s by Lorenz C. Leopold at the request of John Sloop, however, notes that the first runs with liquid hydrogen/liquid fluorine took place on 6 May 1958, when a water-cooled engine with a converging showerhead injector was tested.[19] Although the run logs are not clear on this point, tests of fluorine may actually have predated those with liquid oxygen as the oxidizer. The run log for 23 May 1958 noted a serious accident when 800 pounds of fluorine were released into the atmosphere. Although no injuries resulted from this accident, it dampened the researchers' enthusiasm for fluorine. "Just when you thought you had everything figured out and licked, it up and bit you," Frank Kutina recalled, referring to the problems with fluorine.[20] Interestingly, according to the Leopold document, the first liquid hydrogen/liquid oxygen run using gaseous fluorine ignition did not occur until 20 January 1959.

The running of such a large facility forced the group to specialize. Research scientist William Tomazic recalled that before the RETF was completed, rocket engineers were jacks-of-all-trades. They conceived the test, took part in the design of the equipment to be used, installed it in the test cell, and ran the tests assisted by one or two mechanics. June Bahan Szucs, the group's first secretary, recalled the informality of that period, including the practical jokes they played on each other. Occasionally, she would be asked to push the start button to initiate a test. She soon realized that she was so honored particularly when there was a good chance the test would fail. When a test article blew up, engineers reacted with mock horror, though such "mishaps" were not uncommon.[21]

After the RETF was built, more formal procedures were necessary. As the tests became more numerous and complicated, engineers began to define themselves as research scientists, hardware designers, or operations people.[22] Although the research scientists conceived the test programs, the operations engineers, familiar with every idiosyncrasy of the Rocket Engine Test Facility, were in charge of the actual operation of the facility during a test. The operations engineers designed and built the test article and set it up in the test cell. Electrical support engineers were responsible for the instrumentation, data acquisition, and controls, including the cameras for high-speed photography, the closed circuit TV systems, and the lighting.

The relationship between the research engineers, the hardware designers, and the operations engineers is another example of how the RETF functioned as the locus for the transformation of ideas into hardware. A propulsion laboratory is what some historians of technology call a "community of practice."[23] It includes scientists, engineers, hardware designers, and technicians with different talents and specialties, each of whom brings different knowledge and experience to the task. Although the research engineer usually takes the lead in the design of a test program, he or she depends on the laboratory's hardware designers and operations engineers to configure the test article and instrumentation to yield useful data.

The laboratory was particularly fortunate to have a gifted hardware designer on its staff. Infatuated with rockets since the age of 10, Ohioan George Repas had spent his high school years building and firing them. After graduating from college, he headed to the White Sands Missile Range, and then to Cape Canaveral to test Pershing rockets. Out of a job at the end of the Pershing Program in 1963, he returned to Ohio to work under Frank Kutina at the RETF. Repas spent his career at Lewis designing hardware. At this time, fabrication was seldom contracted out because of the in-house talent available. Repas played a key role in the interface between research engineer and the design of an experiment destined for the RETF. Injectors became his specialty. A research engineer would discuss the goals of a particular research program he had in mind, and Repas would produce an injector design and supervise its fabrication in the shop.[24]

Between five and eight operations engineers, or "ops engineers" as they called themselves, actually ran the facility. They kept detailed notes on each test and minutes of their section meetings. These handwritten logs convey a sense of adventure and camaraderie. Noted in the section meeting log for 5 December 1957: "We carry ball on plans for hdware [sic] & testing, assembling, etc. . . . We have to use this down time to think out future steps on future engines. How to handle it, etc. How to analyze data. Regenerative [cooling] program is virgin data."[25]

As RETF operations became more formal, researchers were required to submit a "research requirements document" to the operations group located in the test facility. The operations people would then meet with the research engineers to determine whether the requested tests were realistic and collaborate in designing a series of test runs. Depending on the complexity of the tests, preparation of the data system and other instrumentation could take up to three months. Setting up a new test program took considerable creativity on the part of the operations engineers. Neal Wingenfeld, an electrical engineer who worked in the RETF operations for more than 30 years, said, "We actually take their ideas and we design the hardware. We have to make it work."[26] Valves and pipes might need to be resized and new electronic control systems designed. Douglas

Bewley, an electrical engineer hired by NASA in the 1980s, described the creative exchange that took place between research and operations engineers:

> A researcher would come with the research requirements document to the operations group, and the operations mechanical side and the operations electrical side would get together, meet with the research engineer, go over his parameters, see what was realistic and what was not realistic, and try to come up with unique ways of doing things. If he really needed to get some data, we would invent ways of being able to do the test so we could get the data. At which point, we would then build the facility up, and depending upon how complicated the program was, it could take anywhere from two months to three months to build up a test program down at the facility and also prepare the data system and all the instrumentation.[27]

Research engineer Ned Hannum emphasized that "absolutely nothing of value could have ever happened without the operations people being able to perform their tricks to simulate environments and make things happen in a cost-effective, quick way." He marveled at how the operations people could take down one engine and install another one on the test stand the same day. Occasionally, they were able to test three complete "build-ups" the same night.[28]

Once the test parameters had been agreed upon, however, the research engineer had to step back and let the operations people run the test. The operations people brooked no interference from the research engineers during a test run. The log for 22 April 1958 noted, "Guy who runs engine is [the] boss—nobody should flip [the] switch until told to do so."[29]

Operations engineers thought nothing of working all day to set up a test, check the systems, and calibrate the instrumentation. They would then work through the night running tests, which would end about eight the next morning. In the 1960s—before the Center

Operations engineers in the control room, located in Building 100, monitor early rocket tests, 1957. Video cameras in the test cell transmitted images during rocket firing. Instruments recorded pressures and other data that was later analyzed by the research engineers in charge of a particular test program.
NASA C-45021.

resumed its work on air-breathing engines—the RETF was among the most heavily used test facilities at Lewis.

When the facility was ready for a test, the South-40 area was cleared. The test conductor, about four operations people, and five or six research engineers assembled in the control room in Building 100 to monitor the test. Although the test conductor always kept his hand poised over the abort or "panic button" during a test, the limitations of reaction time almost precluded human control. Tests usually lasted less than a minute and were too dangerous to be directly observed. In the late 1950s and early 1960s, cam-operated electric switches were used to sequence the test runs. These timers were replaced in the 1970s with programmable logic controllers (PLCs). The PLCs were actually programmable computers designed to function like the old electromechanical timers. The new technolo-

gy was vastly more accurate and versatile. PLCs could be programmed to open the valves and actuators controlling fuel, oxidant,
and ignition sequence with much more precision and regularity. If a
problem developed, test engineers could manually override the PLCs.

The recording of such data as engine temperatures and pressures
and other performance criteria happened automatically. In the late
1950s, analog data was recorded on magnetic tape and strip charts.
Oscillating graphic linear recorders, manufactured by the Honeywell
Company of Minneapolis, provided temperature, pressure, and flow
rates simultaneously. This data had to be reduced rapidly by the
research engineers, since it was used to define the parameters for the
next test and to assess the condition of the hardware.

During a test, data from the rocket engine being tested was fed to
the Instrument Room next to the test cell. In the Instrument Room,
the Transient Analog Data Acquisition System (TRADAR) digitized,
formatted, and transmitted the data up to the control room in
Building 100. It was then transmitted as a file to the central computer building, located in the 10-foot-by-10-foot wind supersonic tunnel office building.

In the mid-1960s, a Digital Data Acquisition System (DDAS) was
introduced. Digital computing dramatically increased the speed and
accuracy of test data reduction. The RETF was one of the first facilities at Lewis to adopt the new digital technology. Interestingly, during this transition, linear recorders continued to be used in the control room because they provided immediate feedback. When a new
Research and Analysis Center (RAC) was built at Lewis in 1979, the
RETF relied on IBM 3033 TSS and Cray 1-S computers, as well as
DEC VAX clusters to reduce the data and transmit it back to the
control room where it could be analyzed. Computers provided engineers with close to real-time data acquisition.[30]

While facility hardware and software evolved over the life of the
facility, the test sequence remained virtually unchanged from the
1960s until it closed in the 1990s. Prior to the start of an engine
fueled with liquid hydrogen/liquid oxygen, the igniter system was

pressurized and test-fired. Then the liquid hydrogen line leading into the injector was prechilled with liquid helium. At the same time, the propellant tanks were pressurized. The PLC-controlled procedures were divided into carefully choreographed periods called "zones" for accomplishing specific tasks.

During the first 15-second period, the engine was readied for firing. The test conductor or engineer-in-charge pushed the start button to activate the data control systems that automatically calibrated the instruments and started recording pressure and temperature data on strip charts. Gaseous nitrogen, used to purge the propellant lines, and water started to flow into the scrubber. After 11 seconds, the program initiated a 2-second liquid oxygen flow to cool the oxidizer line and injector. At 13 seconds gaseous nitrogen purges cleared the engine of the liquid oxygen.

The 3- to 5-second period before the test engine was fired was the most critical time. First the igniter was turned on. Then two propellant fire valves automatically opened, allowing liquid hydrogen and liquid oxygen to flow into the rocket engine where they ignited. Chamber pressure was continuously monitored by safety systems after the fire valves were opened. The "start transient" was an extremely critical period because propellant flows and chamber pressure could oscillate. Propellant and combustion parameters were monitored continuously after combustion, and if they strayed above or below a predetermined point, the computers would terminate the test. The test conductor also kept his hand near the abort button. After a lapse of between two and half and four seconds, the fire valves were closed and the engine shut down. Then the propellants were flushed from the lines beyond the fire valves, the purges restarted, and finally, facility shutdown initiated. During the final 120-second period, engine shutdown was completed and the area rendered safe for inspection. Calibrations were taken again, the scrubber water turned off, and all data systems turned off.[31]

At any time during a test, if engineers monitoring data acquisition in the control room noted abnormal propellant and chamber pressures, the engineer-in-charge could immediately abort the test by pressing the abort button. However, it was more likely that the computers would sense a problem and shut down the test. During an abort, propellant fire valves automatically slammed shut. The shut-off valves on the two propellant tanks also automatically closed, and the prime vent valves, located in the line between the tank shut-off valves and the fire valves, opened to vent any propellants trapped in the line. Slamming the tank valves shut prevented any unburned propellants from escaping into the test facility where they could cause an explosion. Any unplanned event required a full investigation to determine the causes before testing could be resumed.

RETF provided rocket engine designers with data that increased their design options. The facility could be adapted to a range of test environments and test types to support research in rocket processes and materials. RETF engineers published the results of their test programs as NASA research notes, memoranda, or reports. They also presented papers at meetings of national organizations, such as the American Rocket Society, and later the American Institute of Aeronautics and Astronautics (AIAA), created by the merger of the American Rocket Society and the Institute of the Aeronautical Sciences in 1963.

The first RETF research report appeared as a NASA Memorandum in March 1959. The authors of "Experimental Performance of Gaseous Hydrogen and Liquid Oxygen in Uncooled 20,000-Pound-Thrust Rocket Engines" were Edward Rothenberg, Frank Kutina, and George Kinney.[32] They sang the praises of hydrogen's properties as a rocket fuel. "Its low viscosity, high specific heat, and low critical pressure make it an excellent regenerative coolant," they said. "Hydrogen enters the combustion chamber above its low critical temperature after cooling the thrust chamber. This gives the advantage of achieving complete combustion in a smaller chamber, since time is not required for vaporization of the fuel." One of the drawbacks of liquid hydrogen fuel, however, was hydrogen's low density, which meant that relatively large tanks would be required, thereby increasing the overall weight of the rocket. It should be noted that this report referred to the structural weight of a *missile* rather than a *launch vehicle*. The report pointed out that prior to this report the limit of tests of the hydrogen/oxygen combination had been 3,000 pounds of thrust, referencing Aerojet Engineering Corporation test data. This made the testing of a 20,000-pound engine in the new Rocket Engine Test Facility at Lewis an extraordinary accomplishment for its time.

A second report by William Tomazic, Edward Bartoo, and James Rollbuhler, called "Experiments with Hydrogen and Oxygen in Regenerative Engines at Chamber Pressures from 100 to 300 Pounds per Square Inch Absolute," was issued in April 1960.[33] It described the use of liquid hydrogen for cooling and the damaging effects of both high- and low-frequency oscillations on the engine performance. Like many rocket research reports of this era, this report was classified.

These reports reflect the strong research culture that produced the RETF—a facility whose design evolved out of rocket expertise developed over the previous decade. Knowledge generated through rocket tests in the RETF and communicated through NACA research reports would contribute indirectly to the development of liquid hydrogen rocket engines like the RL10 and the J-2 by industry. With the formation of NASA in 1958, however, came a change in emphasis. The space race with the Soviet Union forced research engineers to focus on immediate needs of the space program. They temporarily suspended their interest in long-range problems to investigate problems of the RL10 engine, which they extensively tested in the laboratory's Propulsion Systems Laboratory. At the same time, the RETF was called into service to tackle the problem of combustion instability in engines for the multi-staged Saturn V and the proposed mammoth engine for the Nova rocket.

NOTES

1. Dawson, *Engines and Innovation,* 158.

2. Sloop, *Liquid Hydrogen,* 92-3.

3. For a detailed explanation, see Robert C. Stewart, "Rocket Engine Test Facility—GRC Building No. 206," (Washington, DC: National Parks Service, HAER No. OH-124-C, 2003).

4. [unknown author], "Considerations affecting layout of Rocket facility at S-40 location, Drawing CDS 10308," 21 June 1954, Box 14, History, Folder 1 of 2, RETF Record Group, GRC.

5. [unknown author], "Remodeling of Rocket Engine Research Facility, Test Cell Building, project C-1537," 19 May 1959, Box 14, History, Folder 1 of 2, RETF Record Group, GRC.

6. See Scrubber memos, 1952–53, Box 1, Folder 40, RETF Record Group, GRC.

7. Frank Kutina interview, 26 March 2004.

8. See P. M. Ordin and T. W. Reynolds, "Design of Cooling System S-40 Rocket Facility,"[no date]; Paul M. Ordin, "Treatment of Waste Water from S-40 Rocket Facility," 2 March 1955; and "Progress report on the experimental investigation of the removal of hydrogen fluoride from the exhaust of a 100-pound thrust liquid-fluorine liquid-ammonia rocket," 1 May 1952, Box 14, History, Folder 1 of 2, RETF Record Group, GRC.

9. "New Engineers Systems Procedures Manual," [no date], Box 14, History, Folder 1 of 2, RETF Record Group, GRC.

10. [unknown author],"Outline for Specifications-S-40 Rocket Facility," 8 Nov. 1954, Box 14, History, Folder 1 of 2, RETF Record Group, GRC.

11. William Tomazic, comments on manuscript, 5 March 2004.

12. Frank Kutina, comments on manuscript, 9 April 2004.

13. Larry Leopold, "R.E.T.F. Safety Procedure," 20 Nov. 1964, Safety File Items, Box 15, RETF Record Group, GRC.

14. Ibid.

15. Run log 1, 25 Feb. 1957-5 March 1961, no page numbers, Box 18, RETF Record Group, GRC.

16. Kutina interview.

17. Run log 1.

18. Eugene Krawczonek interview, 23 Jan. 2003.

19. Lorenz C. Leopold, "S-40-Rocket Engine Test Facility-LeRC," 4 Dec. 1973, attached to letter dated 5 Nov. 1973 from John Sloop to Edmond Jonash, confirming research details on the history of hydrogen used as propulsion fuel, Box 14, History, Folder 2 of 2, RETF Record Group, GRC.

20. Kutina interview.

21. June Bahan Szucs interview, 25 Oct. 2002.

22. William Tomazic interview, 23 Feb. 2003.

23. This expression was coined by Edward W. Constant II in "Scientific Theory and Technological Testability: Science, Dynamometers, and Water Turbines in the 19th Century," *Technology and Culture* 24 (1983): 183-98.

24. George Repas interview, 23 Jan. 2003.

25. Rocket Engine Performance Section Meeting Log 1, 3 July 1957 to 25 Oct. 1959. Both the RETF Section Meeting Logs and Run Logs are located in Box 18, RETF Record Group, GRC.

26. Interview with Neal Wingenfeld and Doug Bewley, 23 Jan. 2003.

27. Ibid.

28. Ned Hannum interview, 22 Jan. 2003.

29. Rocket Engine Performance Section Meeting Log 1.

30. For a more detailed description, see Robert C. Stewart, "Rocket Engine Test Facility—GRC Building 100," (Washington, DC: National Parks Service, HAER No. OH-124-D, 2003); also Wingenfeld and Bewley interview, 23 Jan. 2003.

31. The preceding description was adapted from several memos written by George Repas after an explosion of the TRW pintle engine in October 1993.

These memos can be found referenced as "South 40 Mishap," Box 11, RETF Record Group, GRC.

32. NASA Memo 4-8-59E. Declassified October 1964.

33. NASA TM X-253. Declassified July 1965.

Combustion Instability and Other Apollo Era Challenges, 1960s

AFTER THE NACA BECAME THE nucleus of the new space agency, Hugh Dryden, former NACA Director and now Deputy Administrator of NASA, called Abe Silverstein to Headquarters, where he was placed in charge of the Office of Space Flight Programs.[1] Silverstein headed this office for four years, helping to shape the new agency and taking charge of early planning for the Mercury and Apollo Programs. Silverstein asked rocket researchers John Sloop and Adelbert Tischler to assist him at Headquarters. One of their most important tasks was to help define the new agency's launch vehicle needs, particularly the choice of fuel for the upper stages of Saturn, the rocket that carried the Apollo astronauts to the Moon.[2] Silverstein, Sloop, and Tischler agreed on liquid hydrogen as the ideal fuel for upper stage launch vehicles, but disagreed on the choice of oxidizer. "You see," Tischler recounted in a recent interview, "Sloop was the one that really wanted the higher performance [of fluorine] and when I got to Washington I decided, 'Heck, we're talking manned space flight here . . . We're not going to deal with fluorine.' You don't make that kind of decision out of nothing. By that time, I had the experience to say, 'I know about this stuff.'"[3]

In December 1959, shortly before the formal transfer of the von Braun group at the Army Ballistic Missile Agency to NASA, Silverstein chaired an important committee to decide the upper stages of the Saturn vehicle. In his book, *Liquid Hydrogen as a Propulsion Fuel,* John Sloop described how Silverstein persuaded von Braun

NASA press conference 12 April 1961, held after the flight of Soviet astronaut Yuri Gagarin, the first person to fly in space and the first to orbit the Earth. Left to right: Robert Seamans, Jr., Associate Administrator; Hugh Dryden, Deputy Administrator; James Webb, NASA Administrator; and Abe Silverstein, Director of Space Flight Programs. Silverstein, former Chief of Research at Lewis and later its director, championed the use of liquid hydrogen in the upper stages of the Saturn rocket.
Great Images of NASA, http://grin.hq.nasa.gov. 61-Webb-2.

of the feasibility of liquid hydrogen. The Huntsville team had planned to use RP-1 with liquid oxygen in Saturn's second stage and was willing to consider the hydrogen-fuelled RL10 engine for the third. Von Braun was not convinced, however, that the RL10 could be developed in time. Silverstein argued that the Saturn rocket required the additional power that only liquid hydrogen could provide in all of its upper stages. He presented the von Braun team with an analysis of Saturn configurations prepared by former Lewis researcher Eldon Hall. After a week of heated debate, the von Braun team capitulated. The decision to use liquid hydrogen in the upper stages of Saturn resulted in a new contract between Rocketdyne and NASA for a second liquid hydrogen engine, the J-2, managed by Marshall Space Flight Center. The selection of liquid hydrogen in

Saturn's upper stages proved to be among the space program's key technical decisions. Liquid hydrogen in the Saturn rocket would provide NASA with an unmistakable edge in the race to the Moon.[4]

LIQUID HYDROGEN EXPERTISE

Injectors, the area of the engine in which propellants are mixed prior to ignition, had long been the focus of experiments at Lewis because of their complexity and the central role they played in successful rocket engine design. By the mid-1950s, research engineers at NASA Lewis had developed a liquid hydrogen/liquid oxygen injector with a perforated face, called a showerhead. This injector, laboriously fabricated by Edward Baehr, Chief of the Fabrication Division, required elaborate machining. Through testing in some of the smaller rocket test cells, it was discovered that propellant mixing could be facilitated by lightly compressing a group of propellant tubes into a bundle. One propellant was sent through the tubes, while the other was sent through the interstices between the tubes. While the "tube bundle" produced high performance, the tubes tended to burn and melt at the injector face. "To promote better cooling—*controlled* cooling—we came up with the concentric tube concept," Frank Kutina recalled. "We now had just two tubes, one around the other, with one propellant going through the annulus and the other propellant going through the tube."[5] By using this configuration, it was now possible to control the cooling of the face of the injector.

Pratt & Whitney adopted this design for the RL10 engine with one important modification. While the concentric tubes greatly improved cooling, hot gases still caused the injector's faceplate to warp. To solve this problem, the company replaced the solid metal face of the Lewis injector with a porous material called Rigi-Mesh that contributed to atomization of the propellants. The new injector resembled a large, shallow mesh colander that allowed gaseous hydrogen to filter through its holes, cooling the injector face, and thereby reducing thermal stresses.[6]

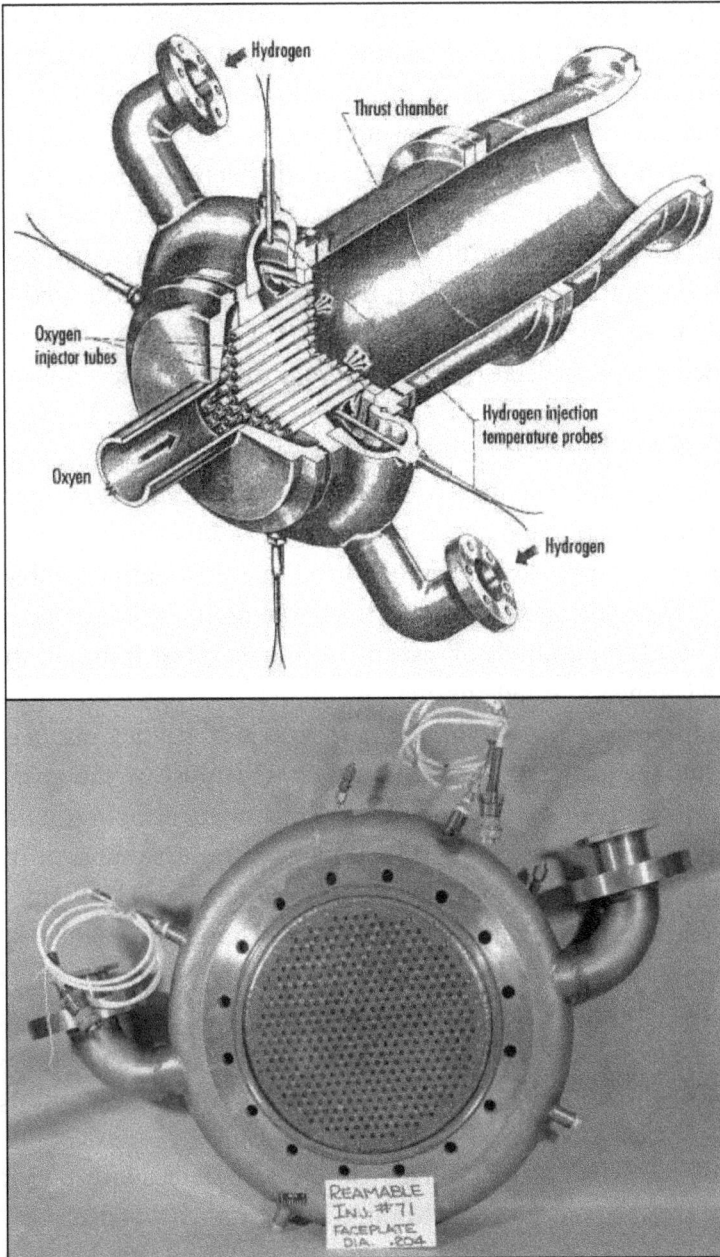

The concentric tube injector developed by members of the rocket section at Lewis in the 1950s. NASA received a patent for the injector in 1962 that is now used in most liquid hydrogen rocket engine designs.
NASA CD8209; NASA C-70277.

However, Pratt & Whitney's intention to patent the concentric tube injector developed at Lewis raised the hackles of NASA management because other companies using this technology would be obliged to license the patent, thereby increasing the cost of rocket engines purchased by NASA. Since the idea had already been used in rocket test models built by hardware designer Samuel Stein for wind tunnel tests, he filed a patent for the invention.[7] After its successful use in the RL10 engine, the concentric tube injector was incorporated into Rocketdyne's J-2. A memo generated after a meeting in May 1963 with Rocketdyne officials proudly announced that the concentric tube injector, developed and tested by the rocket group at Lewis during the pre-NASA years, "has been adopted 100%."[8] It has since become standard in the design of liquid hydrogen injectors, including the Shuttle main engine.

Lewis engineers also became involved in investigating problems related to the ignition system of the RL10 engine. In late 1960 and early 1961, after hundreds of tests in a horizontal test stand, Pratt & Whitney engineers tested the new engine in their new vertical test stand. Rather than the smooth start they expected, they discovered that in the vertical test stand, gravity caused most of the gaseous oxygen to flow out of the combustion chamber, delaying ignition just long enough for the propellants flowing into the exhaust diffuser to explode.[9] RETF engineers found that by squirting a small amount of fluorine into the center of the injector, the engine would start immediately. The "South 40 Run Log" for 3 October 1961 triumphantly reported, "This ended this series of runs with 108 successful starts total!"[10] This solution, however, was never used. Ultimately, the company solved the problem simply by adding a gaseous oxygen valve.

Use of the RETF for problems closely associated with development or trouble-shooting accelerated after the 1961 decision by President John F. Kennedy to land human beings on the Moon before the end of the decade. Researchers at Lewis found themselves swept up in the national emergency created by the perceived superiority of Soviet rocket engines.

COMBUSTION INSTABILITY

Most early rocket engines were plagued by combustion instability, but Rocketdyne's first-stage F-1 engine problems proved especially intractable.[11] The F-1 burned traditional propellants—RP-1 with liquid oxygen as the oxidizer. Its enormous thrust requirements of 1,500,000 pounds made it difficult to scale up from smaller rocket engine prototypes. In June 1962, the explosion of the F-1 engine within half a second after ignition was blamed on combustion instability. The problem placed the entire Apollo Program in jeopardy. Combustion instability was also a problem in the J-2 liquid hydrogen engine used to power the upper stages of the Saturn rocket. Bruce Lundin, Walter Dankhoff, Ward Wilcox, Fred Wilcox, Carl Schueller, Irving Johnsen, and Melvin Hartman from Lewis were among those who helped Rocketdyne tackle the myriad problems associated with the use of liquid hydrogen.[12] Luigi Crocco of Italy, director of the Princeton University Guggenheim Jet Propulsion Center between 1949 and 1968 and a prominent authority on combustion instability, played an important role in this effort.

Combustion instability occurred during the so-called steady-state thrust period of firing. Instead of smooth combustion, pressure changes within the rocket engine caused very rapid oscillations of the gases in the center of the combustion chamber. Low- frequency oscillations damaged delicate instrumentation, reduced thrust, and in some cases caused engines to explode. Dealing with the phenomenon of screaming or screech, caused by high-frequency acoustic waves, was even more daunting. Theory suggested that as these oscillating combustion gases moved to the wall of the combustion chamber, they scrubbed off the boundary layer of slower moving gases next to the wall that under normal conditions provided protection from the extreme heat of combustion. Loss of the boundary layer increased heat transfer, causing the chamber wall to melt down.[13] Compelling as this theory appeared, it was too general to be used as a guide in the design of the injector and thrust chamber. Experience, painstakingly acquired through testing, would prove a better teacher.

A technician inspects a 20,000-pound-thrust rocket engine used to investigate combustion instability, or screech, October 1960. Screech, a serious problem in rocket engines designed for the Apollo program, was investigated in the RETF using sub-scale models of injectors and thrust chambers.
NASA C-54595.

Marshall engineers, who were managing Saturn rocket development, called for a NASA-wide assault on the problem, with $13 million earmarked for Rocketdyne's combustion instability program. Since Lewis research engineers had investigated this problem in the 1950s, they were asked to participate in meetings between Marshall engineers and the contractor.

At Lewis, an ad hoc committee on combustion instability headed by Richard Priem concluded that the design of the F-1's injector was seriously flawed. Priem, Head of the Rocket Combustion Section at Lewis, commented in a memo to Deputy Director Eugene Manganiello that Rocketdyne was reluctant to base baffle designs (thought to dampen the oscillations) on any one theory of combustion instability, because none of the theories had ever been evaluated in hardware that approached full size. Critical of the Rocketdyne staff's ability to find a viable solution to the problem, he recommended that an experimental group be established at Lewis to evaluate the various theories for predicting screech and to test these theories on full-scale injector hardware. Referring specifically to RETF, he said, "This [Lewis] group could perform tests at the 20,000-pound-thrust level using injector designs based on the various theories."[14] He speculated that a solution to the F-1 engine's stability problem would also benefit other liquid hydrogen engines like the J-2.

Despite Priem's recommendation, apparently the program in the RETF did not start immediately. At this time, Lewis engineers were divided over what role the Center should play in the new space agency. Many were reluctant to abandon fundamental research for the rough-and-tumble arena of development. When Priem attended a meeting on the problem of combustion instability of the J-2 engine at Marshall in December 1962, he again complained that hardware testing was not adequate. Queried by Wernher von Braun, Priem admitted that Lewis Research Center policy precluded using its facilities for engine development. He wrote, "The author [Priem] then repeated the policy statement obtained from Dr. Evvard [Deputy Associate Director of Research], that it was Lewis' interpretation of policy that the Lewis Center would not conduct work directly concerned with the F-1 or other development engines."[15]

The urgency of the Apollo Program apparently led to a reversal of this policy and coincided with the return of Abe Silverstein as director of Lewis Research Center. He insisted that the new rocket engine test facility serve NASA's immediate needs. Speaking of the pressures of the Apollo era, mechanic Edward Krawczonek said, "I recall seeing a memo written by Abe Silverstein that he wanted that facility [RETF] running every day. And if it wasn't running every day, he was going to get people that would get it running every day."[16] The RETF staff tested hundreds of different combustion chamber geometries and injector designs based on competing theories advanced by Crocco and David Harrje (also at Princeton University) and Priem and Donald Guentert of NASA Lewis.[17] Ned Hannum, at that time a young engineer working under William Conrad in the RETF, recalled that the shop floor of the test cell (Building 202) was covered with test hardware. The group ran the facility at a frenetic pace, running tests three to four nights a week. If an engine failed, they rapidly set up another, sometimes running two tests in one night. The group produced many papers that described injector designs based on different geometries.[18]

A paper by John Wanhainen, Harold Parish, and William Conrad, "Effect of Propellant Injection Velocity on Screech in 20,000 Pound Hydrogen-Oxygen Rocket Engine," contained the first full description of the RETF's design and operation.[19] This report evaluated the screech stability characteristics of thirteen concentric tube injectors. The report pointed out that though numerous investigations of the phenomenon had been carried out in both experimental and full-scale liquid hydrogen/oxygen engines, design information was lacking. By isolating one variable, the effect of propellant injection velocity, new design information could be generated, allowing the engine designer to design the shape of the injector with greater confidence.

The RETF proved an ideal facility for combustion instability research. The number and quality of the programs demonstrate the creativity of the researchers and the flexibility of the new facility. Different approaches included adding an acoustic liner for the purpose of damping high-frequency oscillations and determining stability transition limits.[20] Researchers also developed a "reamable" injector that allowed them to systematically enlarge (or ream) the orifices of the injector face to measure the effects of propellant pressure drop. Another injector design used an extended tube to determine the effectiveness of heating the injected hydrogen gas before mixing with oxygen in the combustion chamber. Experiments using a coaxial injector (a type of concentric tube injector) reversed the injection relationship of hydrogen and oxygen to determine if a higher heat transfer rate played a role in reducing combustion instability. Another experiment involved injection of gaseous instead of liquid oxygen into the injector. The group developed small bombs, called pyros, installed prior to the test. The bombs, detonated in the combustion chamber, were used to investigate the stability of a particular hardware configuration. Another approach involved the introduction of high-frequency, high-pressure nitrogen into the combustion chamber to induce a siren-like sound.[21]

These studies of combustion instability are examples of parameter variation, a systematic approach to design utilized by engineers when scientific theory cannot adequately predict the performance of a particular piece of hardware. To quote the technology historian Walter Vincenti, "Experimental parameter variation is used in engineering (and only in engineering) to produce the data needed to *bypass the absence of a useful quantitative theory,* that is, to get on with the engineering job when no accurate or convenient theoretical knowledge is available."[22] Parameter variation was one of the hallmarks of American aeronautical research prior to World War II. Aircraft design in the postwar period became more sophisticated with the application of aerodynamic theories based on fluid mechanics. However, because it was far more difficult to use theory to predict the behavior of gases in the combustion chamber of a turbojet or rocket engine, parameter variation continued to be used in rocket engine research as late as the 1960s.

By transforming ideas into hardware and testing various components in the RETF, rocket engineers could systematically explore the problem of combustion instability and suggest various designs to avoid it. Testing enabled the RETF staff to determine the effect of different variables on design, including chamber shapes and acoustic liners. For example, an "eight-element injector" allowed them to study the combustion instability by varying injector elements. These elements could be readily be removed and replaced with others. One experiment involved an ingenious thrust chamber design that resembled the slide on a trombone. It allowed them to test the effect of different chamber lengths without building complete chambers. A "zoned injector" allowed them to vary the flow rate of propellants in a two-part injector that consisted of a central core and annular zone. A "concentrated pattern injector" was used to determine the effect of concentrating the propellant flow near the center of injector. In addition to injectors built in-house, RETF engineers evaluated about 18 to 20 injector designs manufactured by different companies.[23]

In 1963 Leland F. Belew, Chief of the Engine Management Office at Marshall Space Flight Center, strongly supported the idea of an exchange of monthly and quarterly progress reports on the F-1, J-2, RL10, and M-1 Engine Programs between contractors and NASA Centers "so that useful information will be freely exchanged."[24] At this time the RETF was being used for testing a subscale model of the injector for the M-1 engine. This engine program had been transferred from Marshall to Lewis in October 1962, the same month that Lewis took over management of the Centaur Program.[25]

NASA planned to use a giant rocket called Nova to replace the Saturn rocket, originally thought to be too underpowered to transport the astronauts in a direct shot to the Moon. The first stage of Nova would consist of eight Rocketdyne F-1 engines, each generating 1,500,000 pounds of thrust. Four M-1 engines powering the second stage would generate a tremendous thrust of 1,230,000 pounds each. These engines, designed and manufactured by the Aerojet General Corporation of Sacramento, California, would run on liquid hydrogen/liquid oxygen. A single 200,000-pound-thrust liquid oxygen/liquid hydrogen J-2 engine would power the third stage.[26]

The RETF was ideally suited for generating data that could be used in the design of the M-1 engine. It was anticipated that development of the injector, always the pacing item in the development of a new engine, would be slowed because of the problem of combustion instability. Lewis researchers pointed out that since 1956, the screech problem had plagued most rocket engines of over 10,000 pounds thrust. Because of the lack of experimental test data, designers of injectors for very large rocket engines had to rely on building

Cutaway view of the test cell building where two liquid hydrogen storage tanks outside the test cell are being loaded from a mobile dewar, February 1965. The addition of the vent stack on top of the scrubber helped to decrease air pollution.
NASA C-74453.

and testing successive iterations of each injector prototype. In a paper describing the design philosophy for the development of the M-1 injector, the authors wrote: "Lack of basic knowledge on instability, its prevention, and cure has generally forced injector development along the tortuous path of cut-and-try, with its associated long delays and high costs."[27] Experience with the development of the J-2 and the RL10 engines made it possible to establish design criteria in

Aerial view of the RETF in 1963. The old rocket engine test cells can be seen (lower right), protected by earthen mounds. The woods separated the RETF from residential areas of Brook Park to the south.
NASA C-66584.

1964 that allowed Lewis and Aerojet to agree on a single injector design based on the concentric tube developed at Lewis, now referred to as a coaxial tube injector. To verify the correctness of the design while the full-scale injector was being fabricated, a series of 66 test runs of critical components of the engine was carried out in the RETF. The subscale engine used for these tests consisted of a full-scale concentric tube injector, a scaled-down thrust chamber, and a scaled-down convergent-divergent nozzle. This configuration generated 15,000 pounds of thrust and provided important information about how combustion instability would affect the engine's performance. The test program also produced data used in the design of baffles to dampen the effect of combustion instability.[28]

The combustion instability research program generated new instrumentation for tests using liquid hydrogen. Liquid hydrogen research was so new that manufacturers had not yet developed the specialized instruments Lewis researchers needed for engine tests. Jesse Hall, head of the instrument and computing division at that time, invented an instrument that could measure the high frequencies encountered with the screech phenomenon. George Repas and Neal Wingenfeld worked with the Kistler Instrument Corporation of Buffalo to develop a dynamic high-pressure, high-temperature, water-cooled helium bleed transducer that could be mounted on the rocket engine wall to monitor pressure oscillations. When it proved successful, the transducer became a standard catalogue item, sold throughout the world.[29]

The RETF helped to solve the design problems caused by combustion instability in the 1960s. By the mid-1960s, NASA's liquid hydrogen upper stages had proved their feasibility, and by the end of the decade, the Saturn V would fly astronauts to the Moon. The RETF had played a role in making this feat possible. However, if the RETF had been considered a relatively large facility at the time it was completed, within a few years it was dwarfed by the huge test facilities required for Saturn. Nevertheless, among NASA facilities it was still unique because it remained an experimental facility, dedicated to advancing the rocket engine design. Rocket researchers at Lewis continued to try to anticipate the needs of the rocket propulsion community. In the next decade they would find creative new ways to use the facility for rocket research. In the process they would generate new ideas and invent the hardware to test them.

Notes

1. For a discussion of Lewis Laboratory's transition to space, see the author's *Engines and Innovation,* 158-166.

2. See Dawson and Bowles, *Taming Liquid Hydrogen,* 24.

3. A. O. Tischler interview, 19 Nov. 2002.

4. Sloop, *Liquid Hydrogen,* 230-39; Ernst Stuhlinger, "Enabling Technology for Space Transportation," *The Century of Space Science* (Dordrecht: Kluwer Academic Publishers, 2001), vol. 1, 73-4; Asif A. Siddiqi, Challenge to Apollo: *The Soviet Union and the Space Race, 1945–1974* (Washington, DC: NASA Special Publication-2000-4408); Siddiqi, 317-18, 840.

5. Frank Kutina interview, 26 March 2004.

6. Joel E. Tucker, "The History of the RL10 Upper-Stage Rocket Engine 1956–1980;" Stephen E. Doyle, ed., *History of Liquid Rocket Engine Development in the United States 1955–1980* (AAS History Series, vol. 13, 1992), 132; Bilstein, Stages to Saturn, 138.

7. Kutina interview. The concentric tube injector was patented 9 June 1964, patent number 3,136,123, LEW number 111. For abstract see http://timeline.grc.nasa.gov.

8. Ward W. Wilcox to Walter F. Dankhoff, Chief Launch Vehicle Propulsion Office, "J-2 Assessment Team Meeting, 24 April 1963," Chemical Rocket Division, Org. Code 9500B, Box 251A, file 21, M-1 Correspondence, 1964 Historical Records, NASA Glenn Records, GRC.

9. On the development of the RL10 engine, see Dick Mulready, *Advanced Engine Development at Pratt & Whitney: The Inside Story of Eight Special Projects, 1946–1971* (Warrendale, PA: Society of Automotive Engineers, 2001), chap. 4, 57-86; Joel E. Tucker, "The History of the RL10 Upper-Stage Rocket Engine, 1956–1980," in Stephen E. Doyle, ed., *History of Liquid Rocket Engine Development in the United States, 1955–1980,* 123-151.

10. Run Log 1, Box 18, RETF Record Group, GRC.

11. On F-1 development, see Bilstein, *Stages to Saturn,* 104-127; Ray A. Williamson, "The Biggest of Them All: Reconsidering the Saturn V," in Roger D. Launius and Dennis R. Jenkins, eds., *To Reach the High Frontier* (Lexington: University of Kentucky Press, 2002), 314-15.

12. Ward W. Wilcox to Walter F. Dankhoff.

13. Ned Hannum interview, 4 Dec. 2002. See also Martin Summerfield, "The Liquid Propellant Rocket Engine," in *High Speed Aerodynamics and Jet Propulsion Engines* (Princeton, Princeton University Press, 1959), vol. 12, 482-490.

14. Richard Priem to Deputy Director, 21 Aug. 1962. Chemical Rocket Division, Org. Code 9500B, Box 251A, file 21, M-1 Correspondence, 1964 Historical Records, NASA Glenn Records, NASA Glenn Research Center, Cleveland, Ohio. Hereafter cited as GRC.

15. Richard Priem to Deputy Director, 18 Dec. 1962, Chemical Rocket Division, Org. Code 9500B, Box 251A, file 21, M-1 Correspondence, 1964 Historical Records, NASA Glenn Records, GRC. On the issue of research vs. development, see discussion by author in *Engines and Innovation: Lewis Laboratory and American Propulsion Technology* (Washington, DC: NASA Special Publication-4306, 1991), 179-182.

16. Edward Krawczonek interview, 23 Jan. 2003.

17. See Richard J. Priem and Donald Guentert, "Combustion Instability Limits Determined by a Nonlinear Theory and a One-Dimensional Model," NASA TN-D-1408, 1962, and Luigi Crocco, D. I. Harrje, and W. A. Sirignano, "Non-linear Aspects of Combustion Instability in Liquid Propellant Rocket Motors," Second Combustion Conference, Interagency Chemical Rocket Propulsion Group, vol. 1, CPIA Publication, No. 105, Applied Physics Lab., Johns Hopkins University, May 1966, 63-105.

18. See, for example, Harry E. Bloomer, John P. Wanhainen, and David W. Vincent, "Chamber Shape Effects on Combustion Instability," NASA-TM-X-52361, 1967.

19. John P. Wanhainen, Harold C. Parish, and E. William Conrad, "Effect of Propellant Injection Velocity on Screech in 20,000 Pound Hydrogen-Oxygen Rocket Engine," NASA TN D-3373, 1966.

20. Bloomer, H. E., Curley, J. K., Vincent, D. W., and Wanhainen, J. P., "Experimental Investigation of Acoustic Liners to Suppress Screech in Hydrogen-oxygen Engines," NASA-TN-D-3822, 1967; E. William Conrad, Harry E. Bloomer, John P. Wanhainen, "Interim Summary of Liquid Rocket Acoustic-Mode-Instability Studies at a Nominal Thrust of 20,000 Pounds," NASA TN D-4968, 1969.

21. Larry Leopold, "R.E.T.F. Safety Procedure," 20 Nov. 1964, Safety File Items, Box 15, RETF Record Group, GRC.

22. Walter Vincenti, *What Engineers Know and How They Know It* (Baltimore: Johns Hopkins University Press, 1990), 161-62. Italics in original.

23. Leopold, "R.E.T.F. Safety Procedure."

24. Leland F. Belew to Walter F. Dankhoff, 8 March 1963, M-1 Records, Box 522, file 523, NASA Glenn Records, GRC.

25. For a history of the Centaur Program, see Dawson and Bowles, *Taming Liquid Hydrogen.*

26. William Conrad to Chief, M-1 Rocket Branch, Aerojet, 20 July 1964, Box 239B, M-1 Records, NASA Glenn Records, GRC. See also E. W. Conrad, J. K. Curley, and J. P. Wanhainen, "Cooled Baffle Development for M-1 Engine Using a Subscale Rocket Engine," NASA-TM-X-1267, 1966. On M-1, see also Hal Taylor, "Flight of M-1 Delayed Three Years," *Missiles and Rockets,* 22 April 1963, 16.

27. Walter F. Dankhoff, Irving A. Johnsen, E. William Conrad, and William A. Tomazic, "M-1 Injector Development—Philosophy and Implementation," NASA TM D-4730, August 1968.

28. Ibid.

29. George Repas and Neal Wingenfeld/Doug Bewley interviews, 23 Jan. 2003.

chapter 4

Clever Approaches to Testing in the Shuttle Era

I N THE EARLY 1970s, Lewis Research Center experienced a wrench- ing reduction in force that reflected a shift in national priorities. Landing American astronauts on the Moon had marked the fulfill- ment of President Kennedy's 1961 pledge. But faced with the Vietnam War and an escalating national debt, Kennedy's successor, President Lyndon B. Johnson, cut NASA's funding. In the 1970s, the administration of Richard M. Nixon continued to look for ways to trim NASA's budget while seeking to sustain public interest and sup- port for the space program. In 1972, after a meeting with NASA Administrator James Fletcher and his Deputy George Low, Nixon announced a plan to develop the Space Shuttle, a reusable vehicle envisioned by NASA planners since the late 1960s. The Space Shuttle, or Space Transportation System, would dramatically change the American initiative in space. Once the Shuttle was operational, NASA planned to phase out expendable launch vehicles like Atlas- Centaur, Titan-Centaur, and Saturn that were considered less eco- nomical than the proposed new reusable vehicle.[1]

Placed in charge of Shuttle development, Marshall Space Flight Center chose liquid hydrogen fuel for the Shuttle's main engine. This decision shows the full acceptance of liquid hydrogen within the American propulsion technology community. For example, Marshall rocket scientist Ernst Stuhlinger wrote:

Hydrogen rocket technology played an absolutely decisive role not only in the Saturn-Apollo Moon Project, but also in the

Shuttle Project that came to life during the 1970s. A new, high performance engine, the Space Shuttle Main Engine (SSME), was developed in a joint effort by the G. C. Marshall Space Flight Center and the Rocketdyne Division of North American Aviation. By that time, hydrogen technology for rocket engines had reached its full maturity.[2]

Nevertheless, if considered a mature technology by the early 1970s, liquid hydrogen still presented unique design challenges.

THE PLUG NOZZLE

With the Shuttle the dominant effort within NASA, again the RETF demonstrated its utility as a relatively low-cost, extremely flexible test facility. Because the main engine of the Shuttle was to be reused many times, it was important to know exactly how often it could be fired before it developed cracks and ultimately failed catastrophically. Since the engine was operated at elevated thrust chamber pressures, hydrogen's temperature extremes placed unusual stresses on the metal of the thrust chamber. To protect the chamber, it was lined with a copper alloy called NARloy-Z, a regeneratively cooled copper alloy with high thermal conductivity. (NARloy-Z was 96 percent copper, 3.5 percent silver, and .5 percent zirconium.) Liquid hydrogen, initially at minus 410 degrees Fahrenheit, flowed through 470 separate channels in the copper jacket, warmed and became gaseous as it entered the thrust chamber. During firing, hot-gas side-wall temperatures in the throat region of the thrust chamber reached more than 1,000 degrees Fahrenheit. After each firing, the walls of the thrust chamber became thinner and thinner until cracks developed. This phenomenon, called low cycle thermal fatigue, had not been studied in rockets prior to the multiple test firings early in the Space Shuttle Main Engine development program.

Because Lewis was a leader in liquid hydrogen research, Marshall contracted with the Center to tackle the problem of low cycle thermal fatigue. In 1975, Lewis research scientist Richard Quentmeyer

The plug nozzle, used to measure distortion caused by low cycle thermal fatigue, is examined by research engineers Phil Masters (left), Al Pavli (center), and Ned Hannum, 1974. NASA C-74-3466

conceived an approach to testing materials that proved both economical and effective. He asked Carl Aukerman, his section head at the time, to devise a small-scale test article that would conserve hydrogen, which is very expensive.[3]

Aukerman came up with the clever idea for a "plug nozzle" rocket thrust chamber. The plug nozzle consisted of an injector, a constant-diameter outer cylinder, and a water-cooled center body, which, in effect, formed an annular rocket combustion chamber.

Gaseous hydrogen and liquid oxygen were selected as propellants because the gases from this combination provided a high heat flux environment for the materials evaluation. The center body, or plug, was contoured to resemble the combustion chamber, throat, and the supersonic sections of a thrust chamber. The plug nozzle was coated with zirconium oxide to protect the surface from the hot combustion gases and to prolong its life. After several hundred thermal cycles, the coating would erode and the plug could again be coated so it could be reused for additional tests. The outer cylinder, or test article, had cooling passages machined into it and was cooled with liquid hydrogen.[4]

Because the throat section of the Shuttle main engine reached temperatures of about 900 to 1,000 degrees Fahrenheit, the group under Quentmeyer experimented with different, very thin thermal barrier coatings to protect the thrust chamber wall. Cylinders were fabricated from various thrust chamber materials, such as pure copper, half-hard AmZirc (a copper-zirconium alloy), NARloy-Z, and even pure silver. The outer cylinders were coated with materials such as zirconium oxide using NiChrome (nickel and chromium) as the bond coat. The thrust chamber was fired, shut off, and fired again in a cyclic fashion. This test procedure provided a high cyclic thermal strain to the cylinder wall, allowing engineers to evaluate the effectiveness of cooling passages and ceramic thermal barrier coatings. Since the configuration used only about one-seventh the amount of hydrogen consumed in testing a conventional thrust chamber, yet provided the same heat flux level, as many as 85 thermal cycles could be achieved from a single tank of liquid hydrogen in the RETF facility, making it possible to test up to 200 thermal cycles in one evening of testing.

Tests showed that the cooling passages for most combustion chamber liner materials would form a doghouse shape. The cooling passage wall on the hot-gas side of the chamber became thinner after repeated thermal cycles. This eventually produced cracks in the cooling passage wall. Quentmeyer called this phenomenon "thermal ratcheting." He asked metallurgist John Kazaroff from the Chemical Rockets Division and stress analysts like Gary Halford and Mike McGaw from the Structures Division to assist in analyzing the problem. Although the program showed that a dramatic increase in chamber life could be achieved with ceramic coatings, they were not widely adopted because of their tendency to flake off. The test program not only confirmed that NARloy-Z was probably the best choice for the Space Shuttle's engine chamber lining, but also showed that the thrust chamber would need to be replaced far more often than originally thought.[5]

This relatively inexpensive test program was complemented by tests of high-pressure, conventional thrust chambers to provide baseline data. To facilitate this test program, a solid-state cyclic events timer was installed in the RETF in 1972. This programmable timer was designed by Larry Madson and Neal Wingenfeld. It had one millisecond resolution and allowed total automatic control of the tests. Another innovation involved a special thermocouple developed by Clarence Wem, used to measure cooling passage rib temperatures. Before this time, it was only possible to estimate rib temperatures analytically. Between 1973 and 1976, the RETF operations staff performed more than 17,000 test firings of different rocket chamber configurations. The results were reported in a series of papers by research engineers John Kazaroff, Robert Jankovsky, and Albert Pavli, among others.[6]

By the mid-1970s, despite the quality of the data made possible by testing in the RETF, the staff feared the facility would be shut down during the radical downsizing of Lewis Research Center. Wayne Thomas, who headed Operations, made sure that Director Bruce Lundin was informed of the RETF's extraordinary contributions to the Shuttle Program. He pointed out that in 1975 alone the RETF facility had carried out rocket tests 122 days out of the 252 days available for testing. This represented a 50 percent utilization rate—a rate that demonstrated extremely efficient and cost-effective running of a test facility.[7]

Test of a 5,000-pound-thrust engine used to investigate low cycle thermal fatigue, 4 October 1975.
NASA C-75-3125.

Operations engineers Larry Leopold (left) and Wayne Thomas inspect an engine installation in RETF. The test stand is draped with a flag in tribute to the nation's bicentennial, 1976. NASA C-76-3447.

Larry Leopold, in charge of the control room systems setup, kept the test programs on track with his scrupulous attention to detail, while test hardware design engineer George Repas played a key role in making sure the 74 test chamber designs for the program were ready on schedule. Electrical engineer Neal Wingenfeld designed the instrumentation, control system, and Digital Data Acquisition System (DDAS). Joseph Nemeth's role was to assemble special parts of the unique hardware required for this program.[8] At the same time, facility mechanics Edward Krawczonek, James Gerold, Wendell White, Kenneth Whitney, and Howard Cobb never allowed the pressure of running tests to interfere with meticulous test preparations. Since the quality of the data is the measure of a test facility, the instrument systems had to be rechecked and recalibrated daily by technicians Joseph Etzler, Cleophas Cotton, and George Mack. Electricians Thomas Schneider and Dennis Munson rose to the challenge of making sure the electrical connections between the control room and the test stand could support different test programs. June Thompson, stationed in the 10-foot-by-10-foot supersonic wind tunnel, worked furiously to keep the computerized data reduction program continuously updated. Wilbur Dodge and others in the Cryogenic Operations Section maintained the facility's bottle farm, keeping the bottles recharged and making sure liquid hydrogen was delivered on a timely basis.

Quentmeyer's group also studied the use of high-aspect-ratio cooling channels in rocket engine liners to reduce wall temperature. They found that these passages dramatically increased the thrust chamber life. Again RETF researchers and RETF hardware designers developed unique test hardware. Cylinders for the plug nozzle test apparatus were fabricated with 400 cooling channels in the throat region, as compared to 72 channels for most of the cylindrical test chambers. Each cylinder had a ten-thousandth-of-an-inch-wide cooling channel and rib. The design lowered the temperature of the chamber from 1,000 degrees Fahrenheit to between 400 and 600 degrees Fahrenheit. This research program proved that the use of high-aspect-ratio cooling channels in rocket engines could dramatically prolong the life of a rocket engine. The outstanding results of the program were reported in important NASA and AIAA papers. Today, all modern rocket manufacturers, including Rocketdyne, Aerojet, Pratt & Whitney, and TRW, use a copper-based alloy with high-aspect-ratio cooling channels in their combustion chamber designs.[9]

The Low Cycle Thermal Fatigue Program ran tests three or four nights a week. About 50 to 100 test cycles could be run before refilling the tank, and one tank might be reloaded two or three times in one night. "It would sound like an awful, awful large steam engine chugging up a hill, because you would hear this boom, boom, boom," said Neal Wingenfeld.[10] When the RETF was planned in the late 1950s, the section of Brook Park adjacent to the South-40 area

Operations engineers Dave Vincent (with notebook) and Larry Leopold perform a check-out in the control room before a test in 1976. The renovated control room now used digital data-recording systems.
NASA C-76-3748.

was sparsely populated. But by the 1970s, there were more houses, and the neighbors on the other side of the fence were increasingly bothered by the loud and annoying noise of rocket tests. Residents of Cedar Point Road would regularly call the fire station and the main gate to complain.

A proposed expansion of nearby Cleveland-Hopkins International Airport also contributed to the uncertainty surrounding the future of the RETF. In July 1977, Lewis Research Center asked the City of Cleveland for funds to study how to relocate the Rocket Engine Research Facility. Eugene Krawczonek, Chief of the Operations Engineering Branch, justified the continued need for a rocket test facility at Lewis. "Verification of engine life, as well as generation of design criteria and new technology to extend thrust chamber life, will require extensive testing at high chamber pressures," he wrote. "In order to properly investigate the improvement of thrust chamber life, laboratory data must be extended to full-scale thrust chambers, which must be tested at the actual conditions of pressure and temperature." His memo recommended the construction of a facility with a high-pressure propellant feed system and an altitude test capability for testing high expansion area ratio nozzles. He thought propellant tankage and gas bottles from the existing RETF could be reused, but recommended the construction of a new gas scrubber and muffler with a simpler, more efficient design for treating waste and rocket noise abatement.[11]

To make the operation more efficient, in 1981 the operations group requested permission to move liquid oxygen and liquid hydrogen propellant storage from the West Area to a location closer to the RETF. The relocation made deliveries of the propellants safer, since the propellants were placed in a protected and isolated area. Previously, stored propellants had to be moved in heavy mobile dewars across the main section of the laboratory, increasing the chance of the loss of liquid hydrogen fuel through boiling-off. Wayne Thomas requested safety permits for the installation of two dewars, one for 18,000 gallons of liquid hydrogen and the other for 2,000 gallons of liquid oxygen, and a gaseous oxygen tube trailer. Careful planning ensured that barricade mounds between the storage tanks were adequate to protect laboratory personnel from accidental exposure to the leaks, spills, fires, and explosions caused by tank rupture.[12] Despite the threat of airport expansion, the upgraded RETF continued to provide excellent test data. The operations engineers, many of whom had spent their careers in the test facility, were ready for new challenges.

Notes

1. Henry C. Dethloff, "The Space Shuttle's First Flight: STS-1," *From Engineering Science to Big Science: The NACA and NASA Collier Trophy Winners,* Pamela E. Mack, ed. (Washington, D.C.: Government Printing Office, 1998), 286; White House Press Secretary, "The White House, Statement by the President," 5 January 1972, Richard M. Nixon Presidential files, NASA Historical Reference Collection, NASA Headquarters, Washington, DC, in "Nixon Approves the Space Shuttle," Roger D. Launius, *NASA: A History of the U. S. Civil Space Program* (Malabar, Florida: Krieger Publishing Company, 1994), 232.

2. Ernst Stuhlinger, "Enabling technology for space transportation," *The Century of Space Science,* vol. 1, (Dordrecht: Kluwer Academic Publishers, 2001), 73-4.

3. Richard Quentmeyer interview, 22 Jan. 2003.

4. Ibid.

5. Ibid.

6. A selection of these papers on low cycle thermal fatigue includes John M. Kazaroff and George A. Repas, "Conventionally Cast and Forged Copper Alloy for High-Heat-Flux Thrust Chambers," NASA TP 2694, 1987; John M. Kazaroff and Robert S. Jankovsky, "Cyclic Hot Firing of Tungsten-Wire-Reinforced, Copper-Lined Thrust Chambers," NASA TM 4214, 1990; John M. Kazaroff, Albert J. Pavli, and Glenn A. Malone, "New Method of Making Advanced Tube-Bundle Rocket Thrust Chambers," NASA TM 103617, 1990; Kenneth J. Kacynski, "A Three-Dimensional Turbulent Heat Transfer Analysis for Advanced Tubular Rocket Thrust Chambers," NASA TM 103293, 1990.

7. "RETF Facility Presentation for the Director," 7 Jan. 1975, Box 6, 373, RETF Record Group, GRC.

8. [unknown author], "Recognition of the Experimental Rocket Chamber Life Program," [c.1977], Box 14, History, Folder 2 of 2, RETF Record Group, GRC.

9. Richard Quentmeyer in comments on text; Richard J. Quentmeyer, "Experimental Fatigue Life Investigation of Cylindrical Thrust Chambers," AIAA Paper 77-893, July 1977 (NASA TM X-73665); Richard Quentmeyer,

H. J. Kasper, and J. M. Kazaroff, "Investigation of the Effect of Ceramic Coatings on Rocket Thrust Chamber Life," AIAA Paper 78-1034, July 1978 (NASA TM-78892); R. P. Hannum, H. J. Kasper, and A. J. Pavli, "Experimental and Theoretical Investigation of Fatigue Life in Reusable Rocket Thrust Chambers, AIAA Paper 76-685, 1976 (NASA TM-X-73665).

10. Interview with Neal Wingenfeld and Doug Bewley, 23 Jan. 2003.

11. Eugene M. Krawczonek, "Criteria for the Proposed Relocated Rocket Engine Test Facility," 27 Dec. 1977, Wayne A. Thomas Records, Box 7, 413, RETF Record Group, GRC.

12. [Wayne A. Thomas], "Liquid Hydrogen and Liquid Oxygen Storage," Wayne A. Thomas Records, Box 3, Folder 197, RETF Record Group, GRC.

chapter 5

Expanding RETF's Capabilities

THE ADVANTAGES OF A RESEARCH FACILITY like the RETF were its flexibility and the relatively low cost of its test programs. However, the RETF could only test rockets at sea level or atmospheric pressure. What was needed was an additional test stand that could replicate the very low pressures of the upper atmosphere and the vacuum of space. This was particularly important in testing designs for upper-stage launch vehicles.

As early as 1971, the Chemical Propulsion Division had dreamed of conducting a comprehensive evaluation of full-scale rocket engine thrust chambers at high chamber pressure (2,000 psia) and high heat fluxes in order to be able to predict how long a thrust chamber could be safely used in the Space Shuttle. The Shuttle operated at a chamber pressure of 3,000 psia. Space planners envisioned advanced launch vehicles with even higher chamber pressures of up to 4,000 psia. Higher chamber pressures saved weight and decreased spacecraft size.

While actual rocket engines use pumps to create high chamber pressure, RETF engineers simulated pumping by using high-pressure gaseous hydrogen and helium to force liquid propellants into the engine. This approach reduced costs and eliminated the difficulties of maintaining large cryogenic pumps. The RETF's original run tanks had a 1,500 psi capability and high-pressure gas storage at 2,400 psi. In 1978 the run tanks were replaced with new ones that had a 5,000 psi rating, and the old piping was upgraded to handle the higher pressures. However, the gas storage tanks had not been replaced because of their expense, and this meant the facility could not be used at its full capacity.

The story of how the laboratory finally obtained the high-pressure storage tanks it needed began in 1973 when John Kazaroff and Larry Leopold of the Space Propulsion and Power Division heard that four surplus high-pressure tanks capable of the 6,000 psi were available as Air Force surplus materiel. The pair immediately flew to Cape Canaveral, Florida, aboard the NASA-5, a turboprop aircraft used to transport Lewis Research Center personnel. They bravely inspected the four giant "bottles" by climbing inside them, carrying lights and crack identification kits containing dyes. They concluded from their inspection that the five-inch-thick steel tanks were basically sound, though they leaked because of improper welding of their stainless-steel lining. Since repair by the manufacturer, Taylor-Forge Company of Paola, Kansas, would be prohibitively expensive, Kazaroff and Leopold decided to have the bottles shipped to Lewis, where they hoped to be able to have them repaired. Five years passed. At last, in 1978, the four 96,000-pound tanks cleared military bureaucracy and arrived at Lewis. A simple process of welding a sealing strip over all the seams of the liner proved effective.[1]

The higher pressures of the new storage vessels supported the new altitude capability for the facility long advocated by research staff. Test Stand B would utilize the same mechanical systems as the original rocket test stand (now referred to as Stand A), a vacuum test cell to test the nozzles of small, low-thrust rocket engines. In requesting funds for facility construction in the early 1980s, Carl Aukerman, head of the Combustion Devices Section, provided the following justification:

> There is a strong need for altitude simulation capability for rocket engine technology development today. All future orbital transfer vehicle scenarios demand rocket engines of extremely high nozzle area ratios to achieve the necessary performance. Nozzle area ratios up to 1000:1 are included in studies with absolutely no technology available to support these concepts. . . . A modifi-

cation of the RETF at Lewis would make a test facility available to all of industry or academia regardless of where the new concepts originated.[2]

Aukerman noted that new technology was needed in two major areas. The first was to determine the level of performance that was actually attainable with large nozzles, since there was virtually no data available for area ratios beyond 100 to 1. Second, design data was needed for unconventional nozzles, particularly on how to integrate them into a vehicle launched from the Shuttle bay.[3]

After Lewis engineers defined the requirements for Stand B, including a design for a water-cooled supersonic diffuser and intercooler by Richard Quentmeyer, they contracted with Sanders &

Research engineer Tammy Smith Zurawski performs a final check of a large 1000:1 nozzle prior to testing in Stand B, 1987.
NASA C-85-401

Thomas, Inc., to provide a detailed design. They purchased the test stand from the Ormond Company, one of the most respected rocket test stand designers in the country. By using the existing on-site nitrogen supply system to power the ejectors, they were able to reduce construction costs. Construction by the Shirmer Construction Company was completed in 1984 at a cost of about $1 million.

Supervising construction of Stand B, while running tests in Stand A, placed heavy demands on the small RETF operations staff. Although they were responsible for additional test programs, their number did not increase. In a memo to Carl Aukerman, Wayne Thomas noted the pressure they were working under during this period: "I am now constantly playing catchup in an attempt to just keep RETF testing." No allowance had been made for time to recover from breakdowns, test failures, and emergency responses—not to mention increasingly onerous paperwork and the need to respond to management requests. "Unfortunately, many jobs are now being done at minimum level quality. Engineering just to get the work done," he warned. "I sure don't like this situation—interferes with my pride in quality work. This has hurt us schedule-wise; hope it never hurts us safety-wise."[4] The staff responded to the Herculean task of switching between the two test stands on successive days, making the RETF again one of the most heavily used test facilities at Lewis.

Operations engineer Hamilton Fernandez-Ortiz contributed greatly to the successful breaking-in of the new facility. He discovered that in order for the oxygen to remain liquid at the injector face, the engine needed to be prechilled just prior to ignition. He worked out how to get just the right amount of prechilling without dumping excessive oxygen into the altitude chamber. Fernandez-Ortiz described with considerable enthusiasm and trepidation the problems they encountered when they tired to hot-fire the facility during the check-out period:

On Tuesday we got the facility ready to hot fire around 3:30 PM but we were troubleshooting almost all the systems as we were pressurizing them up. Leakages were our main problem. Around 5:00 PM I called Central Control Building and they granted us another half an hour of combustion air. Around 5:00 PM we were on our last step before pushing the start button which is bringing up the diffuser coolant water system. A transducer which reads the pressure of the water coming into the system as we open the supply valve was not reading anything. The cable was bad and never fixed. Also we were not getting an indication from the supply valve that was actually open and the transducer reading delta P across the diffuser cooling system was not reading anything either. Time was running up and shutdown was imminent.[5]

He concluded his communication with this comment: "As far as I know we are still on schedule and with enough enthusiasm to finish the program successfully until a stupid decision from above comes down to tell us different."[6]

Aukerman actively sought additional work for the RETF among NASA's propulsion community. With the international space station on NASA's drawing boards, he thought the RETF could contribute to research on its propulsion requirements. "Our run record is unequalled," he wrote. "We have run an average of 2 times per week, every week since the facility was brought on line in 1958! We would be overjoyed at the opportunity to be a part of the Space Station Program and look forward to taking care of any or all of your thruster [small rocket] testing needs." He highlighted the quality and longevity of the operations staff, many of whom had been running tests in the RETF since the 1960s. He emphasized the RETF's versatility. Because of the addition of Stand B, it was now possible, Aukerman wrote, "to test any propellant combination, at any thrust level, at any pressure, with any nozzle area ratio, for any run duration."[7]

By this time, industry rocket designers were using computer modeling to assist them with design. Computer models allowed them to

A technician mounts a nozzle on the test stand of the new altitude facility, 1985. The new stand created a space-like vacuum for testing low-thrust engines. It utilized the RETF's existing propellant feed systems, air, water, electrical and data-recording systems. NASA C-85-402.

predict the effect of changes in a particular design. This reduced industry's dependence on actual rocket testing, but did not eliminate it entirely. If computer modeling allowed them to take shortcuts, eliminating the laborious construction of hundreds of configurations of a particular design, there were limits to the effectiveness of models in predicting performance. Research engineer Al Pavli wrote, "Recent attempts to design nozzles for these new applications have revealed a great uncertainty in the confidence of the computerized models that were previously thought to be adequate. The uncertainty is sufficiently large that meaningful tradeoffs cannot be made and engine design points cannot be focused."[8]

To acquire the needed data, Pavli spearheaded a plan to test high-area-ratio nozzles in Stand B in the late 1980s. The purpose of these tests was to develop optimum vehicle configurations for upper-stage propulsion systems launched from the cargo bay of the Shuttle. The question was how to maximize the performance of rocket nozzles with area ratios approaching 1000:1 (compared to the 57:1 ratio for the nozzle of the RL10).

The first results from tests of high-area-ratio nozzles in Stand B were published in 1987 in three papers.[9] These papers described combustion chamber pressures of 350 psi that produced a boundary layer along the surface of the nozzle. They compared predictions generated by a computer model with actual nozzle conditions during testing. After these tests were completed, the facility was upgraded to achieve combustion chamber pressures of 1,800 to 2,400 psi. The results, reported in 1996, again showed general agreement with the computer-generated predictions, except that the boundary layers were slightly thinner and loss in performance not as great. Thus, total performance was about two percent higher than predicted. Detailed analysis of this comparison along with boundary layer and heat flux measurements and their comparison with the computer model were reported by R. S. Jankovsky, T. D. Smith, and A. J. Pavli in June 1999.[10]

LOX COOLING: A NEW APPROACH TO AN OLD PROBLEM

New instrumentation enabled rocket researchers to return to the old problem of rocket cooling. Since the days of Robert Goddard in the 1920s and 1930s, the practical problems of how to cool a rocket engine have remained central to its design. Typically, the rocket's fuel is used to cool the walls of the thrust chamber. For example, Rocketdyne's 1.5 million-pound-thrust F-1 engine sent RP-1 kerosene-based fuel coursing through the cooling passages prior to firing. This was effective as long as chamber pressures remained relatively low. However, at higher pressures, RP-1 left sooty deposits in the rocket's cooling passages, leading to the possibility of catastrophic failure. As an alternative to RP-1 cooling, cryogenic liquid oxygen (LOX) offered tantalizing possibilities for cooling rocket engines. Interest in LOX cooling increased in the 1980s when single-stage-to-orbit and heavy-lift launch vehicles were contemplated for the space missions of the 1990s. The new vehicles would need more efficient booster engines, and it was recognized that RP-1 propellants were limited in their cooling capacity at high temperatures and pressures.

The RETF proved an especially suitable facility for the investigation of LOX cooling. A paper by Harold G. Price and Philip A. Masters, "Liquid Oxygen Cooling of High Pressure LOX/hydrocarbon Rocket Thrust Chambers," published in June 1986, discussed advanced Earth-to-orbit propulsion technology and an experimental program using RP-1 and LOX as propellants and supercritical LOX as the coolant.[11] Five thrust chambers with identical coolant passage geometries were tested in the RETF. The LOX cooling program, carried out in Stand A of the RETF between 1986 and 1990, featured tests of two different water-cooled chambers vertically mounted on the test stand that fired downward into the exhaust-gas scrubber.[12] Both chambers were equipped with a triplet impinging element injector that provided the best propellant mixing, fuel vaporization, and mass flux distribution. This injector was specially designed for use

The RETF after the $1 million modification to create the new altitude facility (Stand B), 1985. This modification made the RETF even more versatile, allowing engineers to rapidly switch instrumentation and control from one test stand to the other.
NASA C-85-4093

Test of liquid oxygen cooling of a RP-1/Liquid oxygen engine in Stand A, 1983.
NASA C-83-2643

with hydrocarbon fuel. Its outer ring consisted of 24 fuel holes with 24 inner liquid oxygen holes acting as showerheads.

Test hardware was monitored by means of three closed-circuit television cameras and a test cell microphone. Data was recorded on magnetic tape at the same time that a high-speed photographic camera recorded what was happening in the test cell at a rate of 400 frames per second. Data was converted from analog to digital and sent to an IBM 370 computer located in the Research and Analysis Center.

Testing revealed that the thickness of the carbon deposits (soot) along the calorimeter wall could be predicted and varied with oxygen/fuel mixture ratios. Most striking was the discovery that carbon deposits on the wall of the combustion chamber actually provided a thermal barrier coating that reduced thermal strain on the wall.

Research engineer Elizabeth Armstrong made her reputation through her contributions to LOX cooling. After soliciting comments and suggestions from the larger engineering community, she helped devise a test program to provide confidence in liquid oxygen cooling of hydrocarbon-fueled rocket thrust chambers. She was particularly interested in what happened when oxygen leaked through cracks that formed between the cooling passages and the hot-gas sidewall of the combustion chamber, especially in the area between the throat and the injector. Hardware designer George Repas recalled, "There was this terrible fear that if you ever got a crack inside of your engine, then all of a sudden you're dumping oxygen in the engine and it might blow up."[13] After four years of testing in the RETF, Armstrong triumphantly reported, "As a result of the test series, the reason for the failure occurring in the earlier work was determined to be injector anomalies. The LOX leaking through the simulated fatigue cracks did not affect the integrity of the chambers."[14] Although this research was never incorporated into American rocket engine designs, Lewis rocket researchers were both surprised and gratified to learn that LOX cooling was an extremely viable concept and was already being used by the Russians in the RD-170 Energia/Zenit engine.[15]

INDUSTRY COLLABORATION

By the early 1990s, RETF was more than 30 years old, but its useful life was not yet over. Despite a decade of talk that the facility would be closed down to make way for a new airport runway, the inability of the City of Cleveland and Brook Park to reach an agreement allowed the RETF to continue to operate through the 1980s. In 1992, TRW responded to NASA's industry-wide call to develop a powerful but economical rocket engine. TRW had already developed an unusually trouble-free injector for engines that burned storable hypergolic propellants. (Hypergolic propellants spontaneously ignite when they come in contact with an oxidizer.)

The TRW injector, called the coaxial pintle injector, had been tested with both LOX/propane and LOX/RP-1 fuels, but never with liquid oxygen/liquid hydrogen propellants. In engines with conventional injectors, injection of liquid hydrogen typically produced unstable combustion at the high temperatures proposed for this engine. TRW believed that with the coaxial pintle injector their engine would produce stable combustion. Frank Stoddard, senior project engineer for TRW, wrote in a draft for the project requirements document, "If this technology can be successfully demonstrated, it has the potential to significantly lower launch costs for both government and commercial launch vehicle service users."[16]

TRW had investigated test facilities run by the Air Force and Marshall Space Flight Center, but chose the RETF because of its lower costs and the expertise of the staff in testing liquid hydrogen rocket engines. Doug Bewley vividly recalled the initial meeting with six TRW engineers and five Lewis operations engineers and a technician. "They totally went into shock, because they had been all over the country, and they had never seen a facility of this size having only four or five operations engineers," Bewley said. "They were a little leery of bringing their engine in here to test because they didn't know what kind of data we would give them."[17] Bewley admitted that the Lewis operations team also approached the cooperative agreement with TRW with a measure of trepidation because they had never tested such a large engine.

The staff recognized that this was an important program for Lewis Research Center. At a time of continuing cuts in NASA funding, Lewis management had actively campaigned to win the TRW project for the RETF. "Now that we have the test program, it is up to all of us here to unite and perform an excellent job to show Headquarters and Marshall that the right decision was made," Mark Klem, research project engineer, wrote to the staff describing the agreement with TRW. "Everyone should remember when we are working on this project that we are reflecting an image of NASA and an image of Lewis employees and facilities. These images can play a part in Headquarters making decisions on Lewis' future budgets, manpower levels, programs, and facility upgrades."[18]

This new relationship with industry represented an important reorientation of the research activities associated with the RETF. In the 1960s the Center had been reluctant to investigate the problems associated with combustion instability because of its close connection with rocket engines then under development for the Apollo Program. In the 1970s and 1980s research programs had focused on future technology applications. NASA Lewis research engineers had initiated these programs, but presumably little of this research had influenced the design of actual production engines. With technology transfer a priority within NASA in the 1990s, the management of Lewis aggressively sought a collaborative relationship with industry. When asked to participate directly in the development of the pintle engine, the operations staff responded with enthusiasm.

George Repas made trips to Redondo Beach, California, to discuss hardware and operational procedures with TRW's Space and Technology Group. He found TRW cooperative, but emphasized NASA's "responsibility to do this job safely, correctly, and in accord with our South 40 standards."[19] In his view, prudence dictated that NASA be responsible for the final decisions on matters pertaining to safe operation of the facility.

In a memo to Project Coordinator Elizabeth Armstrong, operations engineers Repas, Morgan, Fernandez-Ortiz, and Tom Soldat addressed each of the operational issues that needed to be resolved. One of the concerns was that the lower efficiency of the engine would result in excess hydrogen being injected into the scrubber. They warned that additional torches might be needed for hydrogen burnoff if flow rates exceeded what was considered safe operation. Of the several approaches to cooling the injector to the temperature of liquid hydrogen, the group favored using liquid helium because of

The TRW pintle engine mounted in the RETF test stand, 1993. At a thrust of 40,000 pounds, it was the largest program carried out in the RETF before the facility was shut down in 1995.
NASA C-93-6570

its inherent safety and simplicity.[20] Of all the questions raised, the most serious was whether the scrubber, which had been allowed to deteriorate in anticipation of shutdown, could adequately cool the exhaust. An evaluation noted a number of bad valves in the spray system and recommended they be locked open, repaired, or replaced with new spool pieces.[21]

At this time the staff submitted a "Construction of Facility Modification" proposal for $12.5 million to upgrade and substantially rehabilitate the facility in order to enhance its productivity. The project included a new "Distributed Control System," rehabilitation of Building 100, and other miscellaneous systems, including substantial repairs to the exhaust scrubber/silencer at a cost of $880,000. Because the spray bars in the scrubber were badly corroded, plans called for new, corrosion-resistant, stainless-steel spray bar components to be installed.

TRW sent their chief instrumentation supervisor, engineers, and truckloads of computer equipment to Lewis. When it was found that this equipment did not mesh with the Lewis computer systems, engineers Wingenfeld and Bewley convinced TRW to use the computer and instrumentation systems that were already in place. The company insisted that the run tanks be maintained at a constant pressure and that the pressure losses in the valves and lines connecting the run tanks and the engine be kept to a minimum to simulate the conditions of the turbopump-fed systems for which the engine was being designed. Repas and Fernandez-Ortiz established a procedure to transfer liquid nitrogen into the low pressure LOX tank and pressurized the whole system for the first time in many years. Electrical engineers Bewley and Soldat, called in at the eleventh hour, implemented a new pressure-fed control system that worked flawlessly. Once the tests got underway, TRW's reservations disappeared. They were surprised at the quality of the data. "Great data started coming in," Bewley recalled. "The test was conducted and pulled off on time and on budget, and they were just elated."[22]

The first phase of the program, completed in March 1992, focused on the stability and efficiency of the engine and the durability of the engine's ablative material. Initial tests were so successful, Stoddard wrote, that "our early notion of what needed to be done in the follow-up testing has changed somewhat."[23] In contracting for second-phase testing, he said:

The objective of these tests is to obtain data on the performance of the low cost ablative material, combustion performance and stability in extended firings, long endurance pintle injector operation with cryogens, scalability of pintle-injector engines and the effect of chamber configuration on combustion performance and stability. All these items are important for establishing a data base to support future scale up to very large engine sizes.[24]

This time TRW staff did not insist on bringing a brigade of engineers from California to oversee testing, and they were still amazed that all this could be done with such a small group of people.[25]

TRW wanted the engine hot-fired at full thrust for long durations, accumulating up to 200 seconds before the ablative liner on the engine had to be stripped out and refurbished. In September 1993, during the fifth test run of the engine that evening, the test article blew up. Investigation revealed the cause was a "hard start" in which the propellants failed to be delivered to the combustion chamber in the right timing sequence.[26] In describing the "big bang" to Phil Kramer, Chair of the Area 5 Safety Committee, Repas wrote that immediately after the explosion took place, he hit the abort button. A cloud obscured the video camera for a short while. He actuated the high-volume purges of gaseous nitrogen that flow through the engine to clean out the propellant line. Then the fire was quenched with carbon dioxide and all systems were depressurized and secured. "We waited until all combustible alarms were down to zero," said Repas. "Then we went to the test cell to determine the extent of the damage."[27] It was a great relief to find that no one in the terminal room adjacent to the test cell was injured. "I suspect some hearts were pumping a little faster but everyone was A-OK," he wrote.[28] Despite this setback, the three phases of the TRW test program were considered an outstanding success.

Nevertheless, unbeknownst to the RETF operations staff, who were proceeding with the facility's rehabilitation, the RETF's days were numbered. In 1995 the cities of Cleveland and Brook Park announced they had finally hammered out an agreement that allowed for the extension of one of the airport runways at Cleveland-Hopkins International Airport. Because this runway crossed land that included the site of the RETF, NASA management decided to cancel the rehabilitation project and announced that the facility would close permanently. The TRW tests were completed during the first half of 1995, with the official shutdown taking place on 1 July 1995. The city completed the new runway in 2004.

CONCLUSION: THE RETF AS RESEARCH TOOL

Among NASA's facilities for testing rocket engines, the Rocket Engine Test Facility in Cleveland was unique. Unlike facilities at the Army Ballistic Missile Agency and later the Marshall Space Flight Center, which were used for development, the RETF remained an experimental test facility throughout the almost three decades of its existence. It was designed to carry out research on key components of rocket engines fueled with high-energy cryogenic propellants, especially liquid hydrogen. This purpose was consistent with the research character of NASA's predecessor organization, the National Advisory Committee for Aeronautics.

Like many other NACA facilities, the RETF was designed in-house. It evolved from research on high-energy propellants by members of the rocket group at Lewis in the 1950s. Convinced of the theoretical superiority of liquid hydrogen and liquid fluorine as rocket propellants, they focused on designing experimental hardware to test this assumption. Fluorine proved too hazardous to handle, but emboldened by their experience flying an aircraft on liquid hydrogen fuel, they became liquid hydrogen's advocates. By developing engine components, testing this hardware, and widely disseminating their test data throughout the rocket propulsion community, they helped win acceptance for liquid hydrogen. By the time they had completed the RETF in 1957, the Lewis rocket group, along with engineers at Pratt & Whitney, had become the nation's experts in liquid hydrogen rocket technology. The concentric tube injector, developed and tested at Lewis during this period, represented an important breakthrough in hydrogen-fueled rocket engine design.

The high quality and the innovative nature of the work at Lewis contributed to the decision by NASA to use liquid hydrogen in the upper stages of the Saturn rocket, the launch vehicle that propelled the astronauts to the Moon—a decision that arguably contributed to winning the space race with the Soviet Union. Today liquid hydrogen is routinely used in successful rocket engines manufactured by the United States, France, and Russia.

The RETF's systems for gathering rocket test data continued to evolve in the 1970s and 1980s. In the early years of the Apollo Program, the RETF had a thrust capability of 20,000 pounds. Though later increased to 80,000 pounds, this was a very small thrust capability, relative to the huge thrust requirements for Saturn rocket engines and later the Space Shuttle Main Engine. However, this small thrust capability did not compromise the RETF's utility as a research tool. In fact, testing sub-scale engine components made the facility more economical to run and allowed rocket researchers to test their ideas in hardware of their own design. In the 1960s, when engine designers lacked an adequate theory to explain why thrust chambers blew up when combustion became unstable, NASA engineers used parameter variation to systematically test different thrust chamber hardware geometries in the RETF. Ideas and their relationship to hardware again drove Lewis engineers in the 1970s when they sought to understand the phenomenon of metal fatigue in the Space Shuttle Main Engine, caused by cyclic exposure to liquid hydrogen. The development of the plug nozzle—a clever piece of hardware that was inexpensive to produce and could be reused—yielded data useful in the generation of a new theory called thermal racheting.

Many of the operations staff continued to be assigned to the RETF throughout their careers. This continuity was essential to keeping the complicated facility running, while constantly upgrading its systems for data gathering. The addition in the 1980s of an altitude facility increased the RETF's versatility, allowing the staff to test any propellant combination for any length of time, regardless of thrust level, pressure and nozzle area ratio. Even when used to assist industry directly in rocket engine design, as in the case of TRW in the 1990s, the RETF remained a research tool used to evaluate different hardware configurations prior to selecting the best one for development. Until shut down in 1995, the RETF had been continuously in operation since 1957. Its legacy can be found in the engineering knowledge generated by several generations of research and operations engineers.

Today all traces of the RETF have disappeared. Even the valley and creek bed have been filled in to make way for the new airport runway. Nevertheless, the razing of the RETF cannot detract from its importance as a research tool to advance rocket engine design. Pratt & Whitney, Rocketdyne, Aerojet, and TRW—manufacturers of the nation's rocket engines—all reaped benefits from RETF testing. Some advanced ideas and concepts tried out in the RETF are still waiting for implementation. Because testing was almost always associated with the generation of research reports, this knowledge is not lost. It remains part of the nation's intellectual capital and demonstrates the importance of disinterested in-house government research for advancing technology.

Notes

1. "Bottle Babies Revisit Big Mama," draft of article to be printed in *Lewis News,* [no date], Box 14, History, Folder 2 of 2, RETF Record Group, GRC.

2. "Modification of RETF for Altitude Simulation, CoF for FY 1983," Wayne A. Thomas Records, Box 5, file 344, RETF Record Group, GRC.

3. Ibid.

4. Wayne Thomas to Carl Aukerman, "Estimate of My Workload & Editorial Comments," 31 Jan. 1984, RETF Daily Work Plans, Wayne A. Thomas Records, Box 6, file 359, RETF Record Group, GRC.

5. Hamilton Fernandez-Ortiz, "B Stand Status," 7 Feb. 1991, Hamilton Fernandez-Ortiz Records, Box 10, RETF Record Group, GRC.

6. Ibid.

7. Carl Aukerman to Gerry Barna, "RETF Capabilities for Testing SSP Components," 12 Nov. 1984, RETF Facility Description, Wayne A. Thomas Records, Box 6, file 372, RETF Record Group, GRC.

8. Al Pavli and Tamara Smith, "Program Plan: High-Area-Ratio Nozzle Performance and Nozzle Boundary Layer Study," c. October 1986, Box 7, file 458, RETF Record Group, GRC.

9. A. J. Pavli, K. J. Kacynzski, and T. A. Smith, "Experimental Thrust Performance of a High-Area-Ratio Rocket Nozzle," NASA TP 2720, 1987; K. J. Kacynski, A. J. Pavli, and T. A. Smith, "Experimental Evaluation of Heat Transfer on a 1030:1 Area Ratio Nozzle," NASA TP 2726, 1987; T. A. Smith, A. J. Pavli, and K. J. Kacynski, "Comparison of Theoretical and Experimental Thrust Performance of a 1030:1 Area-Ratio Rocket Nozzle at a Chamber Pressure of 2413 KN/M2 (350 Psia)," NASA TP 2725, 1987.

10. Jankovsky, Robert S., Kazaroff, John M., and Pavli, Albert J.: Experimental Performance of a High-Area-Ratio Rocket Nozzle at High Combustion Chamber Pressure, NASA TP 3576, 1996.
NASA Technical Paper 1999-208522.

11. Harold G. Price and Philip A. Masters, "Liquid Oxygen Cooling of High Pressure LOX/Hydrocarbon Rocket Thrust Chambers," NASA TM 88805, August 1986.

12. Philip A. Masters, Elizabeth S. Armstrong, and Harold G. Price, "High-Pressure Calorimeter Chamber Tests for Liquid Oxygen/Kerosene (LOX/RP1) Rocket Combustion," NASA TP 2862, December 1988.

13. George Repas interview, 23 Jan. 2003.

14. Elizabeth A. Armstrong and Julie A. Schlumberger, "Cooling of Rocket Thrust Chambers with Liquid Oxygen," AIAA-90-2120, 26th Joint Propulsion Conference, 16-18 July 1990, Orlando, FL.

15. Dave Dooling, "Soviets Have Difficulty Exploiting Space Lead," *Space News,* 9-15 July 1990, 7-8.

16. Frank J. Stoddard, "Project Requirements Document," draft, 20 June 1990, Hamilton Fernandez-Ortiz, Notebook 4, Box 10, RETF Record Group, GRC.

17. Neal Wingenfeld and Douglas Bewley interview, 23 Jan. 2003.

18. Mark D. Klem, "TRW 40K Lb Engine Test Kick-off Meeting," 23 Dec. 1992, Hamilton Fernandez-Ortiz, Notebook 4, Box 10, RETF Record Group, GRC.

19. George Repas to All Interested Parties, "Notes on TRW Engine Tests," 2 April 1991, Hamilton Fernandez-Ortiz, Notebook 4, Box 10, RETF Record Group, GRC.

20. "Memorandum from Operations Group to Beth Armstrong," 26 April 1991, Hamilton Fernandez-Ortiz Notebook 4, Box 10, RETF Record Group, GRC.

21. James E. Giuliani and Mark D. Klem, "REFT Scrubber Evaluation for 40K TRW Tests," 16 April 1992, Hamilton Fernandez-Ortiz, Notebook 4, Box 10, RETF Record Group, GRC.

22. Wingenfeld and Bewley interview.

23. Frank J. Stoddard to Lawrence Diehl, 12 Feb. 1992, Hamilton Fernandez-Ortiz, Notebook 4, Box 10, RETF Record Group, GRC.

24. Ibid.

25. Ibid.

26. George Repas to Phil Kramer, Chair, Area 5 Safety Committee, "South 40 Mishap," 23 Sept. 1993, Box 11, RETF Record Group, GRC.

27. George Repas to Phil Kramer, "South 40 Mishap," 13 Oct. 1993, Box 11, file memos from George Repas, RETF Record Group, GRC.

28. Ibid.

Acknowledgments and Sources

THE ROCKET ENGINE TEST FACILITY, located at NASA Glenn Research Center in Cleveland, Ohio, was listed on the National Register of Historic Places in 1984–1985 for its role in facilitating the overall progress of propulsion technology used in NASA missions and programs. In 1995 it was slated for demolition to make way for a new runway at Cleveland-Hopkins International Airport. When federal money is involved in the alteration or demolition of a historic landmark, the National Historic Preservation Act of 1966 requires that its documentation (referred to as "mitigation") meet standards set by the Secretary of the Interior. In addition to documentation of the RETF for the Historic American Engineering Record, this history of the facility was funded as part of the mitigation effort.

Meghan Hays of History Enterprises, Inc. produced a detailed finding aid for RETF documents removed from the facility prior to demolition. She also assisted in interviewing individuals associated with the project. The RETF Record Group she processed contains records kept by the operations engineers for the facility, including a complete set of the run logs. This collection also contains some records of the research programs associated with the RETF. Some division records were also of relevance, such as those related to the M-1 project and LOX cooling.

John Sloop described the rocket research chronicled in the first chapter in *Liquid Hydrogen as a Propulsion Fuel, 1945–1959*. His papers and transcripts of interviews in the NASA Historical Reference Collection in the NASA History Office in Washington, D.C., also proved valuable. I also had access to several boxes of NASA-Glenn records permanently retained by the National Archives and Records Center in Chicago, Illinois. They contained the NACA Research Authorizations and other NACA materials related to rocket research, unavailable to Sloop when he wrote his book.

An even more valuable source for this history was the corporate memory of various people currently or formerly of NASA Glenn who generously shared their knowledge and read various drafts. I am particularly grateful to Neal Wingenfeld. Others included the late Douglas Bewley, William Goette, Ned Hannum, Edward Krawczonek, Eugene Krawczonek, Frank Kutina, Virginia Morrell, Al Pavli, Richard Quentmeyer, the late George Repas, William Rowe, June Szucs, Adelbert Tischler, and William Tomazic. I would like to thank outside reviewers Tony Springer, Roger Launius and Frank Winter for their comments on the manuscript, and NASA Chief Historian Steve Dick and Glenn History Officer Kevin Coleman for their support of the project.

I also benefited from the assessment of relevant records made by Galen Wilson of the National Archives and Records Administration. Janis Dick and other staff of the NASA Glenn Research Center Technical Library assisted me by locating NASA Technical Reports and checking references. Other people instrumental in making this history a reality included Debbie Demaline, Robert Arrighi, and Bonita Smith of Indyne, Inc., Mark Bowles of History Enterprises, Inc., and Stan Popovich of Hardlines Design Company. Last but not least, I am grateful to Robert C. Stewart and Roy A. Hampton III with whom I collaborated on HAER No. OH-124, "Rocket Engine Test Facility."

NACA/NASA Lewis Research Reports on Rocket Engine Research, 1948–1960*

A. THEORETICAL PERFORMANCE

1. Miller, R. O., and Ordin, P. W.: Theoretical Performance of Rocket Propellants Containing Hydrogen, Nitrogen, and Oxygen. NACA RM E8A30, May 1948.

2. Huff, V. N., and Calvert, C. S., Jr.: Charts for the Computation of Equilibrium Composition of Chemical Reactions in the Carbon-Hydrogen-Oxygen-Nitrogen System at Temperatures from 2000° to 5000°K. NACA TN 1653, July 1948.

3. Huff, V. N., Calvert, C. S., and Erdmann, V.: Theoretical Performance of Diborane as a Rocket Fuel. NACA RM E8I17a, January 1949.

4. Morrell, V. E.: Effect of Combustion-Chamber Pressure and Nozzle Expansion Ratio on the Theoretical Performance of Several Rocket Propellant Systems. NACA RM E50C30, May 1950.

5. Gordon, S., and Huff, V. N.: Theoretical Performance of Lithium and Fluorine as a Rocket Propellant. NACA RM E51C01, May 1951.

6. Huff, V. N., Gordon, S., and Morrell, V. E.: General Method and Thermodynamic Tables for Computation of Chemical Reactions. NACA Report 1037, 1951. (Supersedes TN 2113 and TN 2161).

---

* John Sloop, "NACA High Energy Rocket Propellant Research in the Fifties," Panel on Rocketry in the 1950s, AIAA 8th Annual Meeting, 28 Oct. 1971 (reprint), in John Sloop "Propulsion in the 1950s," misc. papers, NASA Historical Reference Collection, NASA History Office, Washington, DC.

7. Huff, V. N., and Gordon, S.: Theoretical Performance of Liquid Ammonia, Hydrazine and Mixture of Liquid Ammonia and Hydrazine as Fuels with Liquid Oxygen Bifluoride as Oxidant for Rocket Engines. 1—Mixture of Liquid Ammonia and Hydrazine. NACA RM E51L11, February 1952.

8. Huff, V. N., and Gordon, S.: Theoretical Performance of Liquid Ammonia, Hydrazine and Mixture of Liquid Ammonia and Hydrazine as Fuels with Liquid Oxygen Bifluoride as Oxidant for Rocket Engines. 11—Hydrazine. NACA RM E52G09, September 1952.

9. Huff, V. N., and Gordon, S.: Theoretical Performance of Liquid Ammonia, Hydrazine and Mixture of Liquid Ammonia and Hydrazine as Fuels with Liquid Oxygen Bifluoride as Oxidant for Rocket Engines. 111—Liquid Ammonia. NACA RM E52H14, October 1952.

10. Gordon, S., and Huff, V. N.: Theoretical Performance of Liquid Hydrogen and Liquid Fluorine as a Rocket Propellant. NACA RM E52L11, February 1953.

11. Gordon, S., and Huff, V. N.: Theoretical Performance of Liquid Ammonia and Liquid Fluorine as a Rocket Propellant. NACA RM E53A26, March 1953.

12. Gordon, S., and Huff, V. N.: Theoretical Performance of Liquid Hydrazine and Liquid Fluorine as a Rocket Propellant. NACA RM E53E12, July 1953.

13. Gordon, S., and Huff, V. N.: Theoretical Performance of Mixtures of Liquid Ammonia and Hydrazine as Fuel with Liquid Fluorine as Oxidant for Rocket Engines. NACA RM E53F08, July 1953.

14. Gordon, S., and Wilkins, R. L.: Theoretical Maximum Performance of Liquid Fluorine-Liquid Oxygen Mixtures with JP-4 Fuel as Rocket Propellants. NACA RM E54H09, October 1954.

15. Gordon, S., and Huff, V. N.: Theoretical Performance of JP-4 Fuel with a 70-Percent-Fluorine—30-Percent-Oxygen Mixture as a Rocket Propellant. 1—Frozen Composition. NACA RM E56A13a, 11 April 1956.

16. Huff, V. N., Fortini, A., and Gordon, S.: Theoretical Performance of JP-4 Fuel and Liquid Oxygen as a Rocket Propellant. 11—Equilibrium Composition. NACA RM E56D23, 7 September 1956.

17. Gordon, S., and Huff, V. N.: Theoretical Performance of JP-4 Fuel with

a 70-30 Mixture of Fluorine and Oxygen Mixture as a Rocket Propellant. 11—Equilibrium Composition. NACA RM E56F04, 2 October 1956.

18. Huff, V. N., and Gordon, S.: Theoretical Rocket Performance of JP-4 Fuel with Mixtures of Liquid Ozone and Fluorine. NACA RM E56K14, 28 January 1957.

19. Morrell, Gerald: Rocket Thrust Variation with Foamed Liquid Propellants. NACA RM E56K27, 1957.

20. Tomazic, William A., Schmidt, Harold W., and Tischler, Adelbert O.: Analysis of Fluorine Addition to the Vanguard First Stage. NACA RM E56K28, 24 January 1957.

21. Fortini, A., and Huff, V. N.: Theoretical Performance of Liquid Hydrogen and Liquid Fluorine as a Rocket Propellant for a Chamber Pressure of 600 Pounds Per Square Inch Absolute. NACA RM E56L10a, 25 January 1957.

22. Gordon, S., and Drellishak, K. S.: Theoretical Rocket Performance of JP-4 Fuel with Several Fluorine-Oxygen Mixtures Assuming Frozen Composition. NACA RM E57G16a, 1957.

23. Gordon, S.: Theoretical Rocket Performance of JP-4 with Several Fluorine-Oxygen Mixtures Assuming Equilibrium Composition. NACA RM E57K22, 1958.

24. Gordon, S., and Glueck, A. R.: Theoretical Performance of Liquid Ammonia with Liquid Oxygen as a Rocket Propellant. NACA RM E58A21, 1958.

25. Gordon, S., and Kastner, M. E.: Theoretical Rocket Performance of Liquid Methane with Several Fluorine-Oxygen [sic] Assuming Frozen Composition. NACA RM E58B20, 1958.

26. Gordon, S., and McBride, B., J.: Theoretical Performance of Liquid Hydrogen with Liquid Oxygen as a Rocket Propellant. NASA MEMO 5-21-59E, 1959.

27. Gordon, S., Zeleznik, F. J., and Huff, V. N.: A General Method for Automatic Computation of Equilibrium Compositions and Theoretical Rocket Performance of Propellants. NASA TN D-132, 1959.

28. Zeleznik, F. J., and Gordon, S.: An Analytical Investigation of Three General Methods of Calculating Chemical-Equilibrium Compositions. NASA TN D-473, 1960.

29. King, C. R.: Compilation of Thermodynamic Properties, and Theoretical Rocket Performance of Gaseous Hydrogen. NASA TN D-275, 1960.

B. EXPERIMENTAL PERFORMANCE

30. Ordin, P. M., Miller, R. O., and Diehl, J.: Preliminary Investigation of Hydrazine as a Rocket Fuel. NACA RM E7H21, May 1948.

31. Rowe, W. H., Ordin, P. M., and Diehl, J.: Investigation of the Diborane-Hydrogen Peroxide Propellant Combination. NACA RM E7K07, May 1948.

32. Rowe, W. H., Ordin, P. M., and Diehl, J.: Experimental Investigation of Liquid Diborane-Liquid Oxygen Propellant Combination in 100-Pound-Thrust Rocket Engines. NACA RM E9C11 May 1949.

33. Ordin, P. M., and Miller, R. O.: Experimental Performance of Chlorine Trifluoride-Hydrazine Propellant Combination in 100-Pound-Thrust Rocket Engines. NACA RM E9FO1, August, 1949.

34. Ordin, P. M., Douglass, H. W., and Rowe, W. H.: Investigation of the Liquid Fluorine-Liquid Diborane Propellant Combination in a 100-Pound-Thrust Rocket Engine. NACA RM E51I04, November 1951.

35. Ordin, P. M., Rothenberg, E. A., and Rowe, W. H.: Investigation of Liquid Fluorine and Hydrazine-Ammonia Mixture in 100-Pound-Thrust Rocket Engines. NACA RM E52H22, October 1952.

36. Rothenberg, E. A., and Douglass, H. W.: Investigation of the Liquid Fluorine-Liquid Ammonia Propellant Combination in a 100-Pound-Thrust Rocket Engine. NACA RM E53E08, July 1953.

37. Rothenberg, E. A., and Ordin, P. M.: Preliminary Investigation of the Performance and Starting Characteristics of Liquid Oxygen-Liquid Fluorine Mixtures with JP Fuel. NACA RM E53J20, January 1954.

38. Douglass, H. W.: Experimental Performance of Fluorine-Ammonia in 1000-Pound-Thrust Rocket Engines. NACA RM E54C17, May 1954.

39. Tomazic, W. A., and Kinney, G. R.: Experimental Performance of the Mixed-Oxides-of-Nitrogen—Ammonia Propellant Combination with Several

Injection Methods in a 1000-Pound-Thrust Rocket Engine. NACA RM E55A07, 28 March 1955.

40. Douglass, H. W.: Experimental Performance of Fluorine-Oxygen with JP-4 Fuel in a Rocket Engine. NACA RM E55D27, 7 July 1955.

41. Tomazic, W. A., and Rothenberg, E. A.: Experimental Rocket Performance with 15 Percent Fluorine—85 Percent Oxygen and JP-4. NACA RM E55D29, 29 August 1955.

42. Tomazic, W. A.: Rocket-Engine Throttling. NACA RM E55J20, 9 December 1955.

43. Tomazic, W. A., Kutina, F. J., Jr., and Rothenberg, E. A.: Experimental Performance of 5000-Pound-Thrust Rocket Chamber Using a 20-Percent-Fluorine—80-Percent-Oxygen Mixture with RP-1. NACA RM E57B08, 1957.

44. Hendricks, R. C., Ehlers, R. C., and Humphrey, J. C.: Evaluation of Three Injectors in a 2400-Pound-Thrust Rocket Engine Using Liquid Oxygen and Liquid Ammonia. NACA RM E58B25, 1958.

45. Nored, D. L., and Douglass, H. W.: Performance of a JP-4 Fuel with Fluorine-Oxygen Mixtures in 1000-Pound-Thrust Rocket Engines. NASA RM E58C18, 1958.

46. Hendricks, R. C., Ehlers, R. C., and Graham, R. W.: Evaluation of Injector Principles in a 2400-Pound-Thrust Rocket Engine Using Liquid Oxygen and Liquid Ammonia. NASA MEMO 12-11-58E, 1959.

47. Rollbuhler, R. J., and Tomazic, W. A.: Investigation of Small-Scale Hydrazine-Fluorine Injectors. NASA MEMO 1-23-59E, 1959.

48. Rothenberg, E. A., Kutina, F. J., Jr., and Kinney, G. R.: Experimental Performance of Gaseous Hydrogen and Liquid Oxygen in Uncooled 20,000-Pound-Thrust Rocket Engines. NASA MEMO 4-8-59E, 1959.

49. Fortini, A., Hendriux, C. D., and Huff, V. N.: Experimental Altitude, Performance of JP-4 Fuel and Liquid-Oxygen Rocket Engine with an Area Ratio of 48. NASA 5-14-59E, 1959.

50. Miller, Riley O., and Brown, Dwight D.: Effect of Ozone Addition on Combustion Efficiency of Hydrogen-Liquid-Oxygen Propellant in Small Rockets. NASA MEMO 5-26-59E, 1959.

51. Richmond, R. J.: Experimental Performance of a 200-Pound-Thrust Liquid-Pentaborane-Liquid-Hydrazine Rocket. NASA TM X-58, 1959.

52. Sivo, J. N., and Peters, D. J.: Comparison of Rocket Performance Using Exhaust Diffuser and Conventional Techniques for Altitude Simulation. NASA TM X-100, 1959.

53. Stein, S.: A High-Performance 250-Pound-Thrust Rocket Engine Utilizing Coaxial-Flow Injection of JP-4 Fuel and Liquid Oxygen. NASA TN D-126, 1959.

54. Rollbuhler, R. J., and Tomazic, W. A.: Comparison of Hydrazine-Nitrogen Tetroxide and Hydrazine-Chlorine Trifluoride in Small-Scale Rocket Chambers. NASA TN D-131, 1959.

55. Ciepluch, C. C.: Performance of a Composite Solid Propellant at Simulated High Altitudes. NASA TM X-95, 1959.

56. Douglass, H. W., Hennings, G., and Price, H. G., Jr.: Experimental Performance of Liquid Hydrogen and Liquid Fluorine in Regeneratively Cooled Rocket Engines. NASA TM X-87, 1959.

57. Otto, E. W., and Flage, R. A.: Control of Combustion-Chamber Pressure and Oxidant-Fuel Ratio for a Regeneratively Cooled Hydrogen-Fluorine Rocket Engine. NASA TN D-82, 1959.

58. Fortini, Anthony: Performance Investigation of a Nonpumping Rocket-Ejector System for Altitude Simulation. NASA TN D-257, 1959.

59. Tomazic, W. A., Bartoo, E. R., and Rollbuhler, R. J.: Experiments with Hydrogen and Oxygen in Regenerative Engines at Chamber Pressures from 100 to 300 Pounds per Square Inch Absolute. NASA TM X-253, 1960.

60. Jones, W. L., Aukerman, C. A., and Gibb, J. W.: Experimental Performance of a Hydrogen-Fluorine Rocket Engine at Several Chamber Pressures and Exhaust-Nozzle Expansion Area Ratios. NASA TM X-387, 1960. See also, RA P81E45, JO 2C-653R2.

C. COMBUSTION

61. Bellman, D. R., and Humphrey, J. C.: Photographic Study of Combustion in a Rocket Engine. 1—Variation in Combustion of Liquid Oxygen and Gasoline with Seven Methods of Propellant Injection. NACA RM E8F01, June 1948.

62. Heidmann, M. F., and Humphrey, J. C.: Fluctuations in a Spray Formed by Two Impinging Jets. NACA TN 2349, April 1951.

63. Sloop, J. L., and Morrell, G.: Temperature Survey of the Wake of Two Closely Located Parallel Jets Emerging at Sonic Velocity. NACA RM E9I21, 6 Feb. 1950.

64. Jaffe, L., Cross, B. A., and Daykin, D. R.: An Electromagnetic Flowmeter for Rocket Research. NACA RM E50L12, 6 March 1951.

65. Dalgleish, J. E., and Tischler, A. O.: Experimental Investigation of a Light-weight Rocket Chamber. NACA RM E52L19a, 23 March 1953.

66. Heidmann, M. F., and Priem, R. J.: Application of an Electro-Optical Two-Color Pyrometer to Measurement of Flame Temperature for Liquid Oxygen-Hydrocarbon Propellant Combination. NACA TN 3033, October 1953.

67. Bellman, D. R., Humphrey, J. C., and Male, T.: Photographic Investigation of Combustion in a Two-Dimensional Transparent Rocket Engine. NACA Report No. 1134, 1953. (Supersedes NACA RM E8F01.)

68. Kuhns, P. W.: Determination of Flame Temperatures from 2000° to 3000°K by Microwave Absorption. NACA TN 3254, August 1954.

69. Heidmann, M. F., and Auble, C. M.: Injection Principles from Combustion Studies in a 200-Pound-Thrust Rocket Engine Using Liquid Oxygen and Heptane. NACA RM E55Z2, 7 June 1955.

70. Miller, R. O.: Flame Propagation Limits of Propane and n-Pentane in Oxides of Nitrogen. NACA TN 3520, August 1955.

71. Heidmann, M. F.: Injection Principles for Liquid Oxygen and Heptane Using Two-Element Injectors. NACA RM E56D04, 26 June 1956.

72. Heidmann, M. F., Priem, R. J., and Humphrey, J. C.: A Study of Sprays Formed by Two Impinging Jets. NACA TN 3835, 26 June 1956.

73. Auble, C. M.: Study of Injection Processes for Liquid Oxygen and Gaseous Hydrogen in a 200-Pound-Thrust Rocket Engine. NACA RM E56I25a, 9 January 1957.

74. Heidmann, M. F.: A Study of Injection Process for 15-Percent Fluorine—85 Percent Oxygen and Heptane in a 200-Pound-Thrust Rocket Engine. NACA RM E56J11, 15 January 1957.

75. Heidmann, M. F.: Propellant Vaporization as Criterion for Rocket-Engine Design; Experimental Effect of Fuel Temperature on Liquid-Oxygen—Heptane Performance. NACA RM E56E03, 1957.

76. Neu, R. F.: Injection Principles for Liquid Oxygen and Heptane Using Nine-Element Injectors in an 1800-Pound-Thrust Rocket Engine. NACA RM E57E13, 1957.

77. Priem, R. J.: Propellant Vaporization as a Criterion for Rocket Design; Numerical Calculations of Chamber Length to Vaporize a Single Hydrocarbon Drop. NACA TN 3985, 1957.

78. Priem, R. J.: Propellant Vaporization as a Criterion for Rocket-Engine Design; Calculations Using Various Log-Probability Distributions of Heptane Drops. NACA TN 4098, 1957.

79. Priem, R. J., and Clark, B. J.: Comparison of Injectors with a 200-Pound-Thrust Ammonia-Oxygen Engine. NACA RM E56H01, 1959.

80. Priem, R. J., and Hersch, M.: Effect of Fuel-Orifice Diameter on Performance of Heptane-Oxygen Rocket Engines. NACA RM E56126, 1958.

81. Ingebo, R. D.: Drop-Size Distributions for Impinging-Jet Breakup in Airstreams Simulating the Velocity Conditions in Rocket Combustors. NACA TN 4222, 1958.

82. Heidmann, M. F., and Priem, R. J.: Propellant Vaporization as a Criterion for Rocket Engine Design; Relation Between Percentage of Propellant Vaporized and Engine Performance. NACA TN 4219, 1958.

83. Baker, L., Jr.: Chemical and Physical Factors Affecting Combustion in Fuel–Nitric Acid Systems. NACA RM E58D03, 1958.

84. Baker, L., Jr.: A Study of the Combustion Rates of Hydrocarbon Fuels with Red Fuming Nitric Acid in a Small Rocket Engine. NACA RM E58D03a, 1958.

85. Heidmann, M. F., and Baker, L., Jr.: Combustor Performance with Various Hydrogen-Oxygen Injection Methods in a 200-Pound-Thrust Rocket Engine. NACA RM E58E21, 1958.

86. Priem, R. J.: Propellant Vaporization as a Criterion for Rocket-Engine Design; Calculations of Chamber Length to Vaporize Various Propellants. NACA TN 3883, 1958.

87. Bittker, D. A.: An Analytical Study of Turbulent and Molecular Mixing in Rocket Combustion. NACA TN 4321, 1958.

88. Clark, B. J., Hersch, M., and Priem, R. J.: Propellant Vaporization as a Criterion for Rocket-Engine Design; Experimental Performance, Vaporization, and Heat-Transfer Rates with Various Propellant Combinations. NASA MEMO 12-29-58E, 1958.

89. Heidmann, M. F.: Propellant Vaporization as a Criterion for Rocket-Engine Design; Experimental Effect of Chamber Diameter on Liquid Oxygen—Heptane Performance. NASA TN D-65, 1959.

90. Ingebo, R. D.: Photomicrographic Tracking of Ethanol Drops in a Rocket Chamber Burning Ethanol and Liquid Oxygen. NASA TN D0290, 1960.

91. Foster, H. H., and Heidmann, M. F.: Spatial Characteristics of Water Spray Formed by Two Impinging Jets at Several Jet Velocities in Quiescent Air. NASA TN D-301, 1960.

92. Somogyi, D., and Feiler, C. E.: Liquid-Phase Heat-Release Rates of the Systems Hydrazine—Nitric Acid and Unsymmetrical Dimethylhydrazine—Nitric Acid. NASA TN D-469, 1960.

D. COMBUSTION OSCILLATIONS

93. Tischler, A. O., and Bellman, D. R.: Combustion Instability in an Acid-Heptane Rocket with a Pressurized Gas Propellant Pumping System. TN 3936, May 1953. (Supersedes NACA RM E51G11.)

94. Tischler, A. O., Massa, R. V., and Mantler, R. L.: An Investigation of High-Frequency Combustion Oscillations in Liquid Propellant Rocket Engines. NACA RM E53B27, June 1953.

95. Male, T., Kerslake, W. R., and Tischler, A. O.: Photographic Study of Rotary Screaming and Other Oscillations in a Rocket Engine. NACA RM E53A29, May 1954.

96. Male, T., and Kerslake, W. R.: A Method for Prevention of Screaming in Rocket Engines. NACA RM E5 4F28a, August 1954.

97. Moore, F. K., and Maslen, S. H.: Transverse Oscillations in a Cylindrical Combustion Chamber. NACA TN 3152, October 1954.

98. Pass, I., and Tischler, A. O.: Effect of Fuels on Screaming in 200-Pound-Thrust Liquid-Oxygen Fuel Rocket Engine. NACA RM E56C10, 22 June 1956.

99. Priem, R. J.: Attenuation of Tangential-Pressure Oscillations in a Liquid-Oxygen—n-Heptane Rocket Engine with Longitudinal Fins. NACA RM E56C09, 28 June 1956.

100. Feiler, C. E.: Effect of Fuel Drop Size and Injector Configuration on Screaming in a 200-Pound-Thrust Rocket Engine Using Liquid Oxygen and Heptane. NACA RM E58A20a, 1958.

101. Baker, L., Jr., and Steffen, F. W.: Screaming Tendency of the Gaseous-Hydrogen-Liquid-Oxygen Propellant Combination. NACA RM E58E09, 1958.

102. Hurrell, H. G.: Analysis of Injection-Velocity Effects on Rocket Motor Dynamics and Stability. NASA TR R-43, 1959.

103. Wieber, P. R., and Mickelsen, W. R.: Effect of Transverse Acoustic Oscillations on the Vaporization of a Liquid-Fuel Droplet. NASA TN D-287, 1960.

E. IGNITION DELAY OF ACID – FUELS

104. Miller, R. O.: Low Temperature Ignition-Delay Characteristics of Several Rocket Fuels with Mixed Acid in Modified Open-Cup-Type Apparatus. NACA RM E50H16, October 1950.

105. Miller, R. O.: Ignition Delay Characteristics in Modified Open-Cup Apparatus of Several Fuels with Nitric Acid Oxidants within the Temperature Range 70° to -105°F. NACA RM E51J11, December 1951.

106. Ladanyi, D. J.: Orthotoluidine and Triethylamine in Rocket Engine Applications. NACA RM E 52K19, January 1953.

107. Miller, R. O.: Ignition Delays of Some Nonaromatic Fuels with Low-Freezing Red Fuming Nitric Acid in Temperature Range -40° to -105°F. NACA RM E52K20, January 1953.

108. Miller, R. O., and Ladanyi, D. J.: Ignition Delays of Alkyl Thiophosphite with White and Red Fuming Nitric Acids within Temperature Range 80° to -105°F. NACA RM E52K25, February 1953.

109. Ladanyi, D. J., and Miller, R. O.: Comparison of Ignition Delays of Several Propellant Combinations Obtained with Modified Open-Cup and Small-Scale Rocket Engine Apparatus. NACA RM E53D03, June 1953.

110. Miller, R. O.: Preliminary Appraisal of Ferrocene as an Igniting Agent for JP-4 Fuel and Fuming Nitric Acid. NACA RM E53H21, 25 Aug. 1953.

111. Miller, R. O.: Effects of Nitrogen Tetroxide and Water Concentration on Freezing Point and Ignition Delay of Fuming Nitric Acid. NACA RM E53G31, September 1953.

112. Ladanyi, D. J., and Hennings, G.: Organosphosphorus Compounds in Rocket-Engine Applications. NACA RM E54A26, April 1954.

113. Ladanyi, D. J., Miller, R. O., and Hennings, G.: Ignition Delay Determinations of Furfuryl Alcohol and Mixed Butyl Mercaptans with Various White Fuming Nitric Acids Using Modified Open-Cup and Small-Scale Rocket Engine Apparatus. NACA RM E53E29, February 1955.

114. Feiler, C. E., and Baker, Louis, Jr.: A Study of Fuel-Nitric Acid Reactivity. NACA RM E56A19, 6 April 1956.

115. Ladanyi, D. J.: Effects of Variations in Combustion-Chamber Configuration on Ignition Delay in a 50-Pound-Thrust Rocket. NACA RM E56F22, 5 October 1956.

116. Miller, R. O.: Ignition Delays and Fluid Properties of Several Fuels and Nitric Acid Oxidants in Temperature Range from 70° to -105°F. NACA TN 3884, December 1956.

F. ENGINE STARTING

117. Ladanyi, D. J., Sloop, J. L., Humphrey, J. C., and Morell, G.: Starting of Rocket Engine at Conditions of Simulated Altitude Using Crude Monoethylaniline and Other Fuels with Mixed Acid. NACA RM E50D20, July 1950.

118. Ladanyi, D. J.: Ignition Delay Experiments with Small-Scale Rocket Engine at Simulated Altitude Conditions Using Various Fuels with Nitric Acid Oxidants. NACA RM E51J01, January 1952.

119. Hennings, G., and Morrell, G.: Preliminary Investigation of a Chemical Starting Technique for the Acid-Gasoline Rocket Propellant System. NACA RM E52K21, January 1953.

120. Kinney, G. R., Humphrey, J. C., and Hennings, G.: Rocket Engine Starting with Mixed Oxides of Nitrogen and Liquid Ammonia by Flow-Line Additives. NACA RM E53F05, August 1953.

121. Hennings, G., Ladanyi, D. J., and Enders, J. H.: Ignition of Ammonia and Mixed Oxides of Nitrogen in 200-Pound-Thrust Rocket Engines at 160°F. NACA RM E54C19, May 1954.

122. Hennings, G., and Morrell, G.: Low Temperature Chemical Starting of a 200-Pound-Thrust JP-4 Nitric Acid Rocket Engine Using a Three-Fluid-Propellant Valve. NACA RM E55E04, 30 June 1955.

123. Krebs, R. F.: Effect of Fluid-System Parameters on Starting Flow in a Liquid Rocket. NACA TN 4034, 1957.

124. Morrell, G.: Summary of NACA Research on Ignition Lab of Self-Igniting Fuel—Nitric Acid Propellants. NACA RM E57G19, 1957.

125. Morrell, G., and Ladanyi, D.: Chemical Igniters for Starting Jet Fuel-Nitric Acid Rockets. NACA RM E57G26, 1957.

126. Krebs, R. P., and Hart, C. E.: Analysis of Flow-System Starting Dynamics of Turbopump-Fed Liquid-Propellant Rocket. NASA MEMO 4-21-59E, 1959.

127. Straight, D. M., and Rothenberg, E. A.: Ignition of Hydrogen-Oxygen Rocket Engine with Fluorine. NASA TM X-101, 1959.

128. Wanhainen, J. P., Ross, P. S., and DeWitt, R. L.: Effect of Propellant and Catalyst Bed Temperatures on Thrust Buildup in Several Hydrogen Peroxide Reaction Control Rockets. NASA TN D-480, 1960.

G. COOLING

129. Sloop, J. L., and Kinney, G. R.: Internal Film Cooling of Rocket Nozzles. NACA RM E8A29a, June 1948.

130. Kinney, G. R., and Lidman, W. G.: Investigation of Ceramic, Graphite, and Chrome-Plated Nozzles on Rocket Engine. NACA RM E8L16, March 1949.

131. Kinney, G. R., and Sloop, J. L.: Internal Film Cooling Experiments in 4-Inch Duct with Gas Temperatures to 2000°F. NACA RM E50F19, April 1950.

132. Kinney, G. R., and Abramson, A. E.: Investigation of Annular Liquid Flow with Concurrent Air Flow in Horizontal Tubes. NACA RM E51C13, May 1951.

133. Morrell, G.: Investigation of Internal Film Cooling of 1000-Pound-Thrust Liquid Ammonia-Liquid Oxygen Rocket Engine Combustion Chamber. NACA RM E51E04, July 1951.

134. Kinney, G. R.: Internal Film Cooling Experiments with 2- and 4-Inch Smooth-Surface Tubes and Gas Temperatures to 2000°F. NACA RM E52B20, April 1952.

135. Abramson, A. E.: Investigation of Internal Film Cooling of Exhaust Nozzle of a 1000-Pound-Thrust Liquid Ammonia-Liquid Oxygen Rocket. NACA RM E52C26, June 1952.

136. Kinney, G. R., Abramson, A. E., and Sloop, J. L.: Internal Liquid Film Cooling Experiments with Air-Stream Temperatures to 2000°F in 2- and 4-Inch-Diameter Horizontal Tubes. NACA Report No. 1087, 1952. (Supersedes NACA RM's E50F19, E51C13, and E52B20.)

137. Reese, Bruce A., and Graham, R. W.: Experimental Investigation of Heat-Transfer and Fluid-Friction Characteristics of White Fuming Nitric Acid. Purdue University, May 1954. NACA TN 3181.

138. Tischler, A. O., and Humphrey, J. C.: Regenerative-Cooling Studies in a 5000-Pound-Thrust Liquid-Oxygen—JP-4 Rocket Engine Operated at 600-Pounds-Per-Square Inch Combustion Pressure. NACA RM E56B02, 17 April 1956.

139. Graham, R. W., Guentert, E. C., and Huff, V. N.: A Mach 4 Rocket-Powered Supersonic Tunnel Using Ammonia-Oxygen as Working Fluid. NACA TN 4325, 1958.

140. Weston, K. C., and Lunney, G. S.: Heat-Transfer Measurements on an Air Launched, Blunted Cone-Cylinder Rocket Vehicle to Mach 9.7. NASA TM X-84, 1959.

141. Curren, A. N., Price, H. G., Jr., and Douglass, H. W.: Analysis of Effects of Rocket-Engine Design Parameters on Regenerative-Cooling Capabilities of Several Propellants. NASA TN D-66, 1959.

142. Hatch, J. E., and Papell, S. S.: Use of a Theoretical Flow Model to Correlate Data for Film Cooling or Heating an Adiabatic Wall by

Tangential Injection of Gases of Different Fluid Properties. NASA TN D-130, 1959.

143. Hatch, J. E., Schacht, R. L., Albers, L. U., and Saper, P. G.: Graphical Presentation of Different Solutions for Transient Radial Heat Conduction in Hollow Cylinders with Heat Transfer at the Inner Radius and Finite Slabs with Heat Transfer at One Boundary. NASA TR R-56, 1959.

144. Robbins, W. H.: Analysis of the Transient Radiation Heat Transfer of an Uncooled Rocket Engine Operating Outside Earth's Atmosphere. NASA TN D-62, 1959.

145. Liebert, C. H., Hatch, J. E., and Grant, R. W.: Application of Various Techniques for Determining Local Heat-Transfer Coefficients in a Rocket Engine from Transient Experimental Data. NASA TN D-277, 1960.

146. Neu, R. F.: Comparisons of Localized Heat-Transfer Rates in a Liquid-Oxygen—Heptane Rocket Engine Employing Several Injection Methods and Oxidant-Fuel Ratios. NASA TN D-286, 1960.

147. Papell, S. S.: Effect on Gaseous Film Cooling of Coolant Injection Through Angled Slots and Normal Holes. NASA TN D-299, 1960.

H. PROPELLANT PROPERTIES

148. Feiler, C. E., and Morell, G.: Investigation of Effects of Additives on Storage Properties of Fuming Nitric Acids. NACA RM E52J16, December 1952.

149. Ladanyi, D. J., Miller, R. O., Daro, W., and Feiler, C. E.: Some Fundamental Aspects of Nitric Acid. NACA RM E52J01, January 1953.

150. Sibitt, W. L., St. Clair, C. R., Bump, T. R., Pagerey, P. F., Kern, J. P., and Fyfe, D. W.: Physical Properties of Concentrated Nitric Acid. Purdue University, June 1953. NACA TN 2970.

151. McKeown, A. B., and Belles, F. E.: Vapor Pressures of Concentrated Nitric Acid Solutions in the Composition Range 83 to 97 Percent Nitric Acid, 0 to 6 Percent Nitrogen Dioxide, 0 to 15 Percent Water, and in the Temperature Range 20° to 80° C. NACA RM E53G08, September 1953.

152. Feiler, C. E., and Morrell, G. M.: Investigation of the Effects of Fluoride on Corrosion and 170°F of 2S-0 Aluminum and 347 Stainless Steel in Fuming Nitric Acid. NACA RM E53L17b, February 1954.

153. Gundzik, R. M., and Feiler, C. E.: Investigation of the Corrosion of Metals of Construction by Alternate Exposure to Liquid and Gaseous Fluorine. NACA TN 3333, December 1954.

154. Ordin, P. M.: Transportation of Liquid Fluorine. NACA RM E55I23, 8 November 1955.

155. Price, H. G., Jr., and Douglass, H. W.: Material Compatibility with Gaseous Fluorine. NACA RM E56K21, 23 January 1957.

156. Price, Harold G., Jr., and Douglass, Howard W.: Nonmetallic Material Compatibility with Liquid Fluorine. NACA RM E57G18, 2 October 1957.

157. Schmidt, H. W.: Reaction of Fluorine with Carbon as a Means of Fluorine Disposal. NACA RM E57E02, 1957.

158. Ordin, P. M.: Hydrogen-Oxygen Explosions in Exhaust Ducting. NACA TN 3935, 29 April 1957.

159. Schmidt, H. W.: Compatibility of Metals with Liquid Fluorine at High Pressures and Flow Velocities. NACA RM E58D11, 1958.

160. Spakowski, A. E.: The Thermal Stability of Unsymmetrical Dimethyl-Hydrazine. NASA MEMO 12-13-58E, 1958.

161. Schmidt, H. W.: Design and Operating Criteria for Fluorine Disposal by Reaction with Charcoal. NASA MEMO 1-27-59E, 1959.

162. Slabey, V. A., and Fletcher, Edward A.: Rate of Reaction of Gaseous Fluorine with Water Vapor at 35° C. NACA TN 4374, 1958.

163. Rollbuhler, R. J., Kinney, G. R., and Leopold, L. C.: Field Experiments on Treatment of Fluorine Spills with Water or Soda Ash. NASA TN D-63, 1959.

164. Antoine, A. C., and Seaver, R. E.: Impact Sensitivity of Liquid Ozone. NASA TM X-88, 1959.

I. TURBOPUMPS

165. Biermann, A. E., and Kohl, R. C.: Preliminary Study of a Piston Pump for Cryogenic Fluids. NASA MEMO 3-6-59E, 1959

166. Rohlik, H. E., and Crouse, J. E.: Analytical Investigation of the Effect of Turbopump Design on Gross-Weight Characteristics of a Hydrogen-Propelled Nuclear Rocket. NASA MEMO 5-12-59E, 1959.

167. Wintucky, W. T.: Analytical Comparison of Hydrazine with Primary Propellants as the Turbine Drive Fluid for Hydrogen-Fluorine and Hydrogen-Oxygen Altitude Stage Rockets. NASA TN D-78, 1959.

168. Fenn, D. B., Acker, L. W., and Algranti, J. S.: Flight Operation of a Pump-Fed Liquid-Hydrogen Fuel System. NASA TM X-252, 1960.

169. Evans, D. G., and Crouse, J. E.: Analysis of Topping and Bleed Turbopump Units for Hydrogen-Propelled Nuclear Rockets. NASA TM X-384, 1960.

170. Lewis, G. W., Rysl, E. R., and Hartmann, M. J.: Design and Experimental Performance of a Small Centrifugal Pump for Liquid Hydrogen. NASA TM X-388, 1960.

J. EXHAUST NOZZLES

171. Guentert, E. C., and Neumann, H. E.: Design of Axisymmetric Exhaust Nozzles by Method of Characteristics Incorporating a Variable Isentropic Exponent. NASA TR R-33, 1959.

172. Chiccine, B. G., Valerino, A. S., and Shinn, A. M.: An Experimental Investigation of Base Heating and Rocket Hinge Moments for a Simulated Missile Through a Mach Number Range of 0.8 to 2.0. NASA TM X-82, 1959.

173. Musial, N. T., and Ward, J. J.: Over-expanded Performance of Conical Nozzles with Area Ratios of 6 and 9 with and without Supersonic External Flow. NASA TM X-83, 1959.

174. Lovell, J. C., Samanich, N. E., and Barnett, D. O.: Experimental Performance of Area Ratio 200, 25, and 8 Nozzles on JP-4 Fuel and Liquid-Oxygen Rocket Engine. NASA TM X-382, 1960.

175. Sivo, J. N., Meyer, C. L., and Peters, D. J.: Experimental Evaluation of Rocket Exhaust Diffusers for Altitude Simulation. NASA TN D-298, 1960.

176. Campbell, C. E., and Farley, J. M.: Performance of Several Conical Convergent-Divergent Rocket-Type Exhaust Nozzles. NASA TN D-467, 1960.

177. Farley, John M., and Campbell, Carl E.: Performance of Several Method-of-Characteristics Exhaust Nozzles. NASA TN D-293, 1960.

Research Reports Prepared in Connection with Testing in RETF, 1960–1995[*]

A. COMBUSTION INSTABILITY (SCREECH)

1. Conrad, E. W., Parish, H. C., and Wanhainen, J. P.: Effect of Propellant Injection Velocity on Screech in 20,000-pound Hydrogen-Oxygen Rocket Engine. NASA TN-D-3373, 1966.

2. Conrad, E. W., and Hannum, N. P.: Performance and Screech Characteristics of a Series of 2500-pound-thrust-per-element Injectors for a Liquid-Oxygen Hydrogen Rocket Engine. NASA TM-X-1253, 1966.

3. Bloomer, H. E., Vincent, D. W., and Wanhainen, J. P.: Experimental Investigation of Acoustic Liners to Suppress Screech in Hydrogen-Oxygen Engines. NASA TM-X-52253, 1966.

4. Bloomer, H. E., Curley, J. K., Vincent, D. W., and Wanhainen, J. P.: Experimental Investigation of Acoustic Liners to Suppress Screech in Hydrogen-Oxygen Rockets. NASA TN-D-3822, 1967.

5. Hannum, N. P., Russell, L. M., and Wanhainen, J. P.: Evaluation of Screech Suppression Concepts in a 20,000-pound Thrust Hydrogen-Oxygen Rocket. NASA TM-X-1435, 1967.

6. Hannum, N. P., Russell, L. M., Vincent, D. W., and Conrad, E. W.: Some Injector Element Detail Effects on Screech in Hydrogen-Oxygen Rockets. NASA TM-X-2982, 1967.

---

[*] It should be noted that the list of reports based on RETF testing appended to this study is not inclusive. Since the RETF was a research tool for the generation of engineering knowledge, it was not used in report titles, making it difficult to locate all the relevant reports.

7. Bloomer, H. E., Vincent, D. W., and Wanhainen, J. P.: Chamber Shape Effects on Combustion Instability. NASA TM-X-52361, 1967.

8. Morgan, C. J., and Wanhainen, J. P.: Effect of Chamber Pressure, Flow per Element, and Contraction Ratio on Acoustic-Mode-Instability in Hydrogen-Oxygen Rockets. NASA TN-D-4733, 1968.

9. Hannum, N. P., Salmi, R. J., and Wanhainen, J. P.: Effect of Thrust per Element on Combustion Stability Characteristics of Hydrogen-Oxygen Rocket Engines. NASA-TN-D-4851, 1968.

10. Bloomer, H. E., Goelz, R. R., and Hannum, N. P.: Stabilizing Effects of Several Injector Face Baffle Configurations on Screech in a 20,000-pound-thrust Hydrogen-Oxygen Engine. NASA TN-D-4515, 1968.

11. Bloomer, H. E., Conrad, E. W., Vincent, D. W., and Wanhainen, J. P.: Liquid Rocket Acoustic-mode-instability Studies at a Nominal Thrust of 20,000 pounds. NASA TN-D-4968, 1968.

12. Conrad, E. W., Bloomer, H. E., Wanhainen, J. P., and Vincent, D.: Interim Summary of Liquid Rocket Acoustic-Mode-Instability Studies at a Nominal Thrust of 20,000 Pounds. NASA TN-D-4968, 23 July 1968.

13. Morgan, C. J., and Wanhainen, J. P.: Effect of Injection Element Radial Distribution and Chamber Geometry on Acoustic Mode Instability in a Hydrogen-Oxygen Rocket. NASA TN-D-5375, 1969.

14. Hannum, N. P., Phillips, B., and Russell, L. M.: On the Design of Acoustic Liners for Rocket Engines—Helmholtz Resonators Evaluated with a Rocket Engine. NASA TN-D-5171, 1969.

B. M-1 INJECTOR

15. Conrad, E. W., Wanhainen, J. P., and Curley, J. K.: Cooled Baffle Development for M-1 Engine Using a Subscale Rocket Engine. NASA TM-X-1267, 1967.

16. Dankhoff, W. F., Johnsen, I. A., Conrad, E. W., and Tomazic, W. A.: M-1 Injector Development Philosophy and Implementation. NASA TN D-4730, 1968.

C. COOLING WITH ABLATIVE THRUST CHAMBERS

17. Wanhainen, J. P.: Oxygen-Hydrogen Injector Performance and Compatibility with Ablative Chambers. NASA TM-X-2615, 1972.

18. Hannum, N. P., Roberts, W. E., and Russell, L. M.: Hydrogen Film Cooling of a Small Hydrogen-oxygen Thrust Chamber and its Effect on Erosion Rates of Various Ablative Materials. NASA TP-1898, 1977.

D. LOW CYCLE THERMAL FATIGUE

19. Schacht, R. L., Price, H. G., Jr., and Quentmeyer, R. J.: Effective Thermal Conductivities of Four Metal-Ceramic Composite Coatings in Hydrogen-Oxygen Rocket Firings. NASA TN D-7055, 1972.

20. Price, H. G., Jr., Schacht, R. L., and Quentmeyer, R. J.: Reliability and Effective Thermal Conductivity of Three Metallic-Ceramic Composite Insulating Coatings on Cooled Hydrogen-Oxygen Rockets. NASA TN D-7392, 1973.

21. Hannum, N. P., Kasper, H. J., and Pavli, A. J.: Experimental and Theoretical Investigation of Fatigue Life in Reusable Rocket Thrust Chambers. NASA TM-X-73413, 1976. AIAA Paper 76-685.

22. Quentmeyer, R. J.: Experimental Fatigue Life Investigation of Cylindrical Thrust Chambers. NASA TM X-73665, 1977. AIAA Paper 77-893.

23. Quentmeyer, R. J., Kasper, H. J., and Kazaroff, J. M.: Investigation of the Effect of Ceramic Coatings on Rocket Thrust Chamber Life. NASA TM-78892, 1978. AIAA Paper 78-1034.

24. Hannum, N. P., Quentmeyer, R. J., and Kasper, H. J.: Some Effects of Cyclic Induced Deformation in Rocket Thrust Chambers. NASA TM-79112, 1979. AIAA Paper 79-0911.

25. Hannum, N. P., Quentmeyer, R. J., and Price, H. G., Jr.: Some Effects of Thermal-cycle-induced Deformation in Rocket Thrust Chambers. NASA TP-1834, 1980.

26. Kazaroff, J. M., and Repas, G. A.: Conventionally Cast and Forged Copper Alloy for High-Heat-Flux Thrust Chambers. NASA TP 2694, 1987.

27. Quentmeyer, R. J.: Thrust Chamber Thermal Barrier Coating Techniques. CASI 19890013300, 1989.

28. Quentmeyer, Richard J.: Rocket Combustion Chamber Life-Enhancing Design Concepts. AIAA-90-2116, 1990; NASA CR 185257, 1990.

29. Jankovsky, R. S., and Kazaroff, J. M.: A Life Comparison of Tube and Channel Cooling Passages for Thrust Chambers. NASA TM-103613, 1990.

30. Kazaroff, J. M., and Jankovsky, R. S.: Cyclic Hot Firing of Tungsten-Wire-Reinforced, Copper-Lined Thrust Chambers. NASA TM 4214, 1990.

31. Kazaroff, J. M., Pavli, A. J., and Malone, G. A.: New Method of Making Advanced Tube-Bundle Rocket Thrust Chambers. NASA TM 103617, 1990.

32. Kacynski, K. J.: A Three-Dimensional Turbulent Heat Transfer Analysis for Advanced Tubular Rocket Thrust Chambers. NASA TM 103293, 1990.

33. Kazaroff, J. M., and Pavli, Albert J.: Advanced Tube-Bundle Rocket Thrust Chamber. AIAA-90-2726, 1990; NASA TM 103139, 1990.

34. Kacynski, Kenneth J., Kazaroff, John M., and Jankovsky, Robert S.: A Dual-Cooled Hydrogen-Oxygen Rocket Engine Heat Transfer Analysis. AIAA-91-2211, 1991.

35. Kazaroff, J. M., Jankovsky, R. J. and Pavli, A. J.: Hot Fire Test Results of Subscale Tubular Combustion Chambers. NASA TP 3222, 1992.

36. Pavli, Albert J., Kazaroff, John M., and Jankovsky, Robert S.: Hot Fire Fatigue Testing Results for the Compliant Combustion Chamber. NASA TP 3223, 1992.

37. Kazaroff, J. M., and Pavli, A. J.: Advanced Tube-Bundle Rocket Thrust Chambers, Journal of Propulsion and Power 8 (1992): 786-791.

38. Quentmeyer, Richard J., and Roncace, Elizabeth A.: Hot-Gas-Side Heat Transfer Characteristics of Subscale, Plug-nozzle Rocket Calorimeter Chamber, NASA TP 3380, 1993.

39. Jankovsky, Robert S., Kazaroff, John M., and Pavli, Albert J.: Experimental Performance of a High-Area-Ratio Rocket Nozzle at High Combustion Chamber Pressure, NASA TP 3576, 1996.

40. Jankovsky, Robert S., Smith, Timothy D., and Pavli, Albert J.: High-Area-Ratio Rocket Nozzle at High Combustion Chamber Pressure—Experimental and Analytical Validation. NASA TP 1999-208522, 1999.

G. ROCKET NOZZLES

41. Pavli, A. J., Kacynksi, K. J., and Smith, T.A.: Experimental thrust Performance of a High-Area-Ratio Rocket Nozzle. NASA TP 2720, 1987.

42. Kacynski, K. J., Pavli, A. J., and Smith, T. A.: Experimental Evaluation of Heat Transfer on a 1030:1 Area Ratio Nozzle. NASA TP 2726, 1987.

43. Smith, T. A., Pavli, A. J., and Kacynski, K. J.: Comparison of Theoretical and Experimental Thrust Performance of a 1030:1 Area-Ratio Rocket Nozzle at a Chamber Pressure of 2413 KN/M2 (350 Psia). NASA TP 2725, 1987.

H. ROCKET COOLING

44. Carlile, J. A., and Quentmeyer, R. J.: An Experimental Investigation of High-aspect Ratio Cooling Passages. AIAA-92-3154, 1992; NASA TM 105679, 1992.

45. Carlile, J. A.: An Experimental Investigation of High-aspect Ratio Cooling Passages. NASA TM 105679, 1992.

46. Wadel, M. F.: Validation of High-Aspect Ratio Cooling in a 89 KN (20,000 lb. (sub f) Thrust Combustion Chamber. NASA TM 107270, 1996.

47. Armstrong, E. S.: Test Program to Provide Confidence in Liquid Oxygen Cooling of Hydrocarbon Fueled Rocket Thrust Chambers. NASA TM 88816, 1986.

48. Price, H. G., and Masters, P. A.: Liquid Oxygen Cooling of High Pressure LOX/Hydrocarbon Rocket Thrust Chambers. NASA TM 88805, 1986.

49. Naraghi, M.H.N., and Armstrong, E.S.: Three Dimensional Thermal Analysis of Rocket Thrust Chambers. NASA TM 101973, 1988.

50. Masters, P. A., Armstrong, E. S., and Price H. G.: High-Pressure Calorimeter Chamber Tests for Liquid Oxygen/Kerosene (LOX/RP-1) Rocket Combustion. NASA TP 2862, 1988.

51. Armstrong, E. S.: Liquid Oxygen Cooling of Hydrocarbon Fueled Rocket Thrust Chambers. NASA TM 102113, 1989.

52. Armstrong, E. S., and Schlumberger, J. A. [Carlile]: Cooling of Rocket Thrust Chambers with Liquid Oxygen. NASA TM 103146, 1990. AIAA-90-2120.

I. PINTLE ENGINE

53. Klem, M. D., Wadel, M. F., and Stoddard, J. J.: Results of 178 KN (40,000 lbf) Thrust LOX/LH2 Pintle Injector Engine Tests. 32nd JANNAF Combustion Subcommittee Meeting and Propulsion Engineering Research Center 7th Annual Symposium, vol. 2, CPIA-Publ-631, 1995.

Glossary

F-1:   First-stage engine for Saturn V; manufactured by Rocketdyne, burned RP-1/liquid oxygen, generating 1,500,000 pounds of thrust.

J-2:   Rocket engine for upper stages (S-II and S-VB) of Saturn V rocket; manufactured by Rocketdyne, burned liquid hydrogen/liquid oxygen, generating 200,000 pounds of thrust.

M-1:   Engine for upper stage of the planned super booster Nova; manufactured by the Aerojet-General Corporation, burned liquid hydrogen/liquid oxygen to generate 1,230,000 pounds of thrust. Nova was cancelled in 1966.

psi: a measure of pressure expressed in pounds per square inch

psia: pounds per square inch absolute (atmospheric pressure)

psig: pounds per square inch gauge (pressure above atmospheric)

RL10:   Rocket engine for the Centaur upper stage; manufactured by Pratt & Whitney, burned liquid hydrogen/liquid oxygen, generated 15,000 pounds of thrust. The Centaur was fitted with two RL10s.

RP-1,
or JP-4:   used interchangeably to refer to a high-grade kerosene-based fuel used in rockets and turbojet engines.

V-2:   Nazi German missile of World War II and the world's first large-scale liquid-propellant rocket; the V-2 engine burned ethyl alcohol/liquid oxygen. The Redstone missile, developed at the Army Ballistic Missile Agency, was the descendent of the V-2.

Index

www.ingramcontent.com/pod-product-compliance
Lightning Source LLC
Chambersburg PA
CBHW080544110426
42813CB00006B/1206